BARRON'S

The Leader in Test Preparation

STUDENTS'
#1
CHOICE

SAT® SUBJECT TEST

BIOLOGY E/M

Deborah T. Goldberg, M.S.
AP Biology Teacher
Lawrence High School
Cedarhurst, New York

D1500970

BARRON'S

About the Author

Deborah Goldberg earned her B.S. and M.S. degrees from Long Island University. For fourteen years she did research as an electron microscopist at NYU Medical Center and New York Medical College. From 1984 to 2006, Ms. Goldberg worked at Lawrence High School on Long Island, New York. She taught Chemistry and AP Biology, and also developed a popular course in Forensic Science.

Dedication

I wish to dedicate this book to—
  —My wonderful children, Michael and Sara Boilen, whom I love and cherish.

Acknowledgments

I wish to thank—
  —My husband, Howard Blue, for his constant love and support
  —My sister, Rachel Goldberg, for her mastery of the English language and her willingness to share it
  —Pat Hunter, my editor, for her expert guidance and without whom this manuscript would still be sitting in my office
  —My students at Lawrence High School on Long Island, New York, who make teaching the best job in the world
  —Michele Sandifer, expert copy editor

All inquiries should be addressed to:
Barron's Educational Series, Inc.
250 Wireless Boulevard
Hauppauge, New York 11788
www.barronseduc.com

ISBN-13: 978-0-7641-3519-4 (book)
ISBN-10: 0-7641-3519-8 (book)
ISBN-13: 978-0-7641-9344-6 (book/CD-ROM package)
ISBN-10: 0-7641-9344-9 (book/CD-ROM package)

Library of Congress Control Number: 2006028143

Goldberg, Deborah T.
    SAT subject test biology E/M / Deborah T. Goldberg.
      p. cm.
    Includes index.
    ISBN-13: 978-0-7641-3519-4 (book)
    ISBN-10: 0-7641-3519-8 (book)
    ISBN-13: 978-0-7641-9344-6 (book/CD-ROM package)
    ISBN-10: 0-7641-9344-9 (book/CD-ROM package)
    1. Biology—Examinations—Study guides. 2. SAT (Educational test)—Study guides.
    3. College entrance achievement tests—United States—Study guides. I. Title.

QH316G58 2007
570.76—dc22   2006028143

Printed in the United States of America

9 8 7 6 5 4 3

Contents

Contents   **v**

Why Should I Buy This Book?

This book includes:

- Extensive subject reviews of every topic required for the Biology SAT, including the most current information about the three-domain system of taxonomy

- One mini-diagnostic test that identifies your strengths and weaknesses

- Two full-length practice tests with questions that mimic the actual SAT Biology test

- Instructions that enable you to calculate your 200 to 800 SAT score

- Over 400 additional practice questions in the subject reviews

- Several questions that are more difficult than the ones appearing on the actual SAT exam as an additional and optional challenge. These are identified with .

WHAT YOU NEED TO KNOW ABOUT THE SAT SUBJECT TEST: BIOLOGY E/M

The Basics

CHAPTER 1

WHAT IS THE PURPOSE OF THE TEST?

The Biology Subject Test measures your knowledge after completing a general college-preparatory high school Biology course. It is independent of any particular textbook or biology course you may have taken. SAT subject tests are required by many colleges you may decide to apply to.

The SAT Subject Tests fall into five subject areas:

TABLE 1.1

SAT Subject Tests	
Subject Area	**Test Title**
English	Literature
History and Social Studies	US History World History
Mathematics	Math Level 1 Math Level 2
Science	Biology E/M Chemistry Physics
Languages	Chinese with Listening French French with Listening German German with Listening Italian Japanese with Listening Korean with Listening Latin Modern Hebrew Spanish Spanish with Listening

WHEN DO I TAKE THE TEST?

Take the test right after the course ends, when the content is still fresh in your mind. That would be in May or June for most of you. Tests are given in October, November, December, January, May, and June.

HOW DO I REGISTER FOR THE TEST?

You may register by using registration forms found inside the *SAT Registration Bulletin* that is available from your high school Guidance Office.

You can also send for a copy of the *SAT Registration Bulletin* by contacting:

College Board SAT Program
P.O. Box 6200
Princeton, NJ 08541-6299

You can also register for the SAT Subject Test online by visiting the College Board's web site at *www.collegeboard.com*.

You may register by telephone only if you have previously registered with ETS (Educational Testing Service). See their web site for the current telephone number.

When you register, you will have to indicate which specific subject tests you plan to take. **You may take one, two, or three on any given testing date**. Your testing fee will vary accordingly.

You may change your mind on the day of the test and select any test offered and available that day, except for the Language Test with Listening, for which you must preregister.

WHAT IS THE FORMAT OF THE TEST?

The Biology E/M Test has two parts: 60 core questions followed by 20 questions in either specialized section—ecology (E) or molecular (M). Every student answers 80 questions—60 core questions plus 20 special section questions. You choose the area in biology for which you feel most prepared.

There are 5 major content areas on the test. Both tests include questions on all 5 topics. However, the E test has more questions about ecology, while the M test has more questions about molecular and cell biology.

The skills and knowledge that you will be tested on include:

1. **Recall.** Knowledge of facts and terminology

2. **Application.** Understanding concepts, applying knowledge to practical situations, and solving problems

3. **Interpretation.** Integrating information to form conclusions

> The metric system is used on this test.

WHICH TEST SHOULD I TAKE?

If you feel more comfortable answering questions about populations and community systems, and interpreting data about ecosystems, choose the Biology-E exam. If you feel that you would do better answering questions about cell structure and interpreting data about enzymes and other molecules, take the Biology-M test. If you are unsure about which one you should take, discuss it with your biology teacher.

You may decide on the day of the test which test you will take. Because the two tests contain common core questions, you may not take both exams on the same day. You must pencil in the appropriate grid for the test you wish, either Biology-E or Biology-M.

Table 1.2 lists the topics covered on each test. It also indicates the percentage of each topic on the E (ecology) and M (molecular) tests.

TABLE 1.2

Comparison of Biology-E and Biology-M

Core Topics	Approximate Percent of E Test	Approximate Percent of M Test
Cell and molecular biology Cell structure and organization, mitosis, photosynthesis cell respiration, enzymes, molecular inheritance, and biochemistry	15	27
Ecology Energy flow and nutrient cycles, populations and communities, ecosystems and biomes	23	13
Classical genetics Meiosis, Mendelian genetics, inheritance patterns	15	20
Organismal biology Structure, function, and development of organisms with emphasis on plants and animals; animal behavior	25	25
Evolution and diversity Origin of life, evidence of evolution, natural selection, patterns of evolution, taxonomy (classification)	22	15

HOW IS THE TEST SCORED?

The test is scored on a scale of 200 to 800, just like the SAT Test. You earn one point for each correct answer. You lose 1/4 of a point for each wrong answer. Questions you omit are simply not counted and neither are questions for which you mark more than one answer.

You are penalized more for a wrong answer than you are for leaving an answer blank. Your raw score is converted to a scaled score that is reported to you and to any colleges you specify.

After each practice test in this book, you will be directed to grade your own test the same way that the College Board will grade your test.

WHAT IS A GOOD SCORE?

A score of 800 is fabulous, and a 750 is great. However, earning a great score will not necessarily ensure that you get into the college of your dreams. Scoring a 600 does not necessarily mean that you will be barred from your favorite school. Many factors go into college admissions. The SAT Subject Test is simply one factor.

Strategies

Using this book will help you prepare for the SAT Subject Test: Biology E/M. However, do not expect to open up this book for the first time on the night before the test and reap any benefits. Instead, you should use this book for several months and do the following:

1. Familiarize yourself with the directions for the SAT Subject Test.

2. Familiarize yourself with the SAT Subject Test answer sheet.

3. Take the mini-diagnostic test in this book.

4. Study all review chapters in this book and answer all the multiple-choice questions at the end of each chapter, concentrating on your weakest areas.

5. Take the first practice test in this book in a quiet room. Do it **in one sitting**. Do not allow for any distractions. **Do not take any phone calls. Do not talk to anyone. Do not listen to music.**

6. Time yourself. Keep an eye on the time and notice how much of the test you have completed when you are halfway through. Assess if you need to speed up or slow down. The real SAT Subject Test is 1 hour long, so your practice test should also take exactly 1 hour.

> Many students have told me that they found that they were much more nervous during the actual test than when they took the practice test at home. When I asked them under what conditions they took the practice test, they said, "I was listening to some music and took a break in the middle." **Wrong! Wrong! Wrong!** You must mimic the test conditions *exactly* when you practice, or you will not do as well as you could on the test.

7. When you are finished, calculate your grade according to the detailed instructions.

8. Evaluate your results on the practice test.

   Did you have enough time?
   Did you leave too many or not enough questions out?
   Did you make careless mistakes?
   Are there topics you need to study more?

9. After you have evaluated your results on the first practice test, take the second under the same conditions. After you take the second test, evaluate it the same way you evaluated the first one.

10. YOU ARE NOW READY FOR THE SAT SUBJECT TEST IN BIOLOGY.

WHAT TYPES OF QUESTIONS CAN I EXPECT?

1. The Classification Question

This type of question requires little reading and can be answered quickly. Each question has 5 choices, and all the questions refer to those same 5 choices. *A choice may be used once, more than once, or not at all.* Here is an example.

EXAMPLE

Questions 1–3

(A) Chlorophyll
(B) Mitochondria
(C) Smooth endoplasmic reticulum
(D) Nucleolus
(E) Golgi apparatus

1. Has the ability to replicate itself

2. Packages and secretes substances

3. Found in the thylakoid membrane

Answers Explained

1. **(B)** Mitochondria have their own DNA and can replicate themselves independently of the nucleus.

2. **(E)** The Golgi apparatus lies near the nucleus and consists of flattened sacs of membranes stacked next to each other (like a stack of pancakes) and surrounded by vesicles. They modify, store, and package substances produced in the rough endoplasmic reticulum. The Golgi apparatus secretes the substances to other parts of the cell and to the cell surface for export to other cells.

3. **(A)** Chlorophyll is a photosynthetic pigment. It is found in the thylakoid membrane of the grana inside chloroplasts.

2. The Five-Choice Question

The five-choice question is written either as an incomplete statement or as a question. It comes in different varieties. One type is the standard multiple-choice question.

EXAMPLE

4. Thylakoid membranes are most closely related to the

    (A) Sodium-potassium pump
    (B) Krebs cycle
    (C) Human lung
    (D) Inner ear of the human
    (E) Chloroplasts

The answer to question 4 is E. Thylakoid membranes are located in the grana of chloroplasts and are part of the machinery of photosynthesis.

Some five-choice questions contain the words NOT, LEAST, or EXCEPT in capital letters. Read carefully!

EXAMPLE

5. All of the following are found in the grasshopper EXCEPT

    (A) chitinous plates
    (B) capillaries
    (C) veins
    (D) hemocoels
    (E) chitin

The answer to question 5 is B. Grasshoppers have an open circulatory system where blood leaves an artery and seeps through hemocoels (sinuses) to feed body cells before it returns to the heart through a vein. They lack capillaries. Chitinous plates in the grasshopper are located in the gizzard and assist in mechanical digestion. Beside chitinous plates in the gizzard, the exoskeleton is also made of chitin, a carbohydrate.

A special type of five-choice question might be called a multiple-multiple-choice question. It includes a list of 3 or 4 choices labeled by Roman numerals. One or more of these statements may correctly answer the question. You must select from among five lettered choices that follow. Here is one example.

6. ATP is produced during which of the following processes?

    I. Light-dependent reactions
    II. Light-independent reactions
    III. Anaerobic respiration

    (A) I only
    (B) III only
    (C) I and II only
    (D) I and III only
    (E) I, II, and III

The answer to question 6 is D. The light-independent reactions use ATP that is produced in the light-dependent reactions to carry out the Calvin cycle and to produce sugar. Anaerobic respiration produces energy but much less than aerobic respiration.

WHAT TEST-TAKING STRATEGIES SHOULD I USE?

Be Careful

- **Do not make any stray marks on the answer sheet;** the tests are scored by machine.
- You may write anywhere in the test booklet. Use it as scrap paper.
- Check often to see that you are placing the answers in the correct place.
- Be extra careful if you skip a question. Make sure that you also skip a line on the answer sheet. *You cannot waste any time remarking your answer sheet.*

Work at a Steady Pace—Watch the Clock

- The test is 1 hour in length.
- Bring a watch to the exam, and keep an eye on the time.
- Do not spend too much time on any one question. Remember, every question is worth one point, no matter how easy or difficult it is.

It Is OK to Guess

- If you are not sure of an answer but can eliminate at least one choice, then guess.
- If you have no idea about one question, then leave it blank and move on to the next question. Remember to skip the appropriate line on the answer sheet.

No calculators are allowed—computations are easy.

MINI-DIAGNOSTIC TEST

Answer Sheet
MINI-DIAGNOSTIC TEST

1 Ⓐ Ⓑ Ⓒ Ⓓ Ⓔ   9 Ⓐ Ⓑ Ⓒ Ⓓ Ⓔ   17 Ⓐ Ⓑ Ⓒ Ⓓ Ⓔ   25 Ⓐ Ⓑ Ⓒ Ⓓ Ⓔ
2 Ⓐ Ⓑ Ⓒ Ⓓ Ⓔ   10 Ⓐ Ⓑ Ⓒ Ⓓ Ⓔ   18 Ⓐ Ⓑ Ⓒ Ⓓ Ⓔ   26 Ⓐ Ⓑ Ⓒ Ⓓ Ⓔ
3 Ⓐ Ⓑ Ⓒ Ⓓ Ⓔ   11 Ⓐ Ⓑ Ⓒ Ⓓ Ⓔ   19 Ⓐ Ⓑ Ⓒ Ⓓ Ⓔ   27 Ⓐ Ⓑ Ⓒ Ⓓ Ⓔ
4 Ⓐ Ⓑ Ⓒ Ⓓ Ⓔ   12 Ⓐ Ⓑ Ⓒ Ⓓ Ⓔ   20 Ⓐ Ⓑ Ⓒ Ⓓ Ⓔ   28 Ⓐ Ⓑ Ⓒ Ⓓ Ⓔ
5 Ⓐ Ⓑ Ⓒ Ⓓ Ⓔ   13 Ⓐ Ⓑ Ⓒ Ⓓ Ⓔ   21 Ⓐ Ⓑ Ⓒ Ⓓ Ⓔ   29 Ⓐ Ⓑ Ⓒ Ⓓ Ⓔ
6 Ⓐ Ⓑ Ⓒ Ⓓ Ⓔ   14 Ⓐ Ⓑ Ⓒ Ⓓ Ⓔ   22 Ⓐ Ⓑ Ⓒ Ⓓ Ⓔ   30 Ⓐ Ⓑ Ⓒ Ⓓ Ⓔ
7 Ⓐ Ⓑ Ⓒ Ⓓ Ⓔ   15 Ⓐ Ⓑ Ⓒ Ⓓ Ⓔ   23 Ⓐ Ⓑ Ⓒ Ⓓ Ⓔ
8 Ⓐ Ⓑ Ⓒ Ⓓ Ⓔ   16 Ⓐ Ⓑ Ⓒ Ⓓ Ⓔ   24 Ⓐ Ⓑ Ⓒ Ⓓ Ⓔ

E Section

31 Ⓐ Ⓑ Ⓒ Ⓓ Ⓔ   34 Ⓐ Ⓑ Ⓒ Ⓓ Ⓔ   37 Ⓐ Ⓑ Ⓒ Ⓓ Ⓔ   39 Ⓐ Ⓑ Ⓒ Ⓓ Ⓔ
32 Ⓐ Ⓑ Ⓒ Ⓓ Ⓔ   35 Ⓐ Ⓑ Ⓒ Ⓓ Ⓔ   38 Ⓐ Ⓑ Ⓒ Ⓓ Ⓔ   40 Ⓐ Ⓑ Ⓒ Ⓓ Ⓔ
33 Ⓐ Ⓑ Ⓒ Ⓓ Ⓔ   36 Ⓐ Ⓑ Ⓒ Ⓓ Ⓔ

M Section

41 Ⓐ Ⓑ Ⓒ Ⓓ Ⓔ   44 Ⓐ Ⓑ Ⓒ Ⓓ Ⓔ   47 Ⓐ Ⓑ Ⓒ Ⓓ Ⓔ   49 Ⓐ Ⓑ Ⓒ Ⓓ Ⓔ
42 Ⓐ Ⓑ Ⓒ Ⓓ Ⓔ   45 Ⓐ Ⓑ Ⓒ Ⓓ Ⓔ   48 Ⓐ Ⓑ Ⓒ Ⓓ Ⓔ   50 Ⓐ Ⓑ Ⓒ Ⓓ Ⓔ
43 Ⓐ Ⓑ Ⓒ Ⓓ Ⓔ   46 Ⓐ Ⓑ Ⓒ Ⓓ Ⓔ

Mini-diagnostic Test Biology E/M

CHAPTER 3

With Answers and Analysis

Directions: Each of the questions or incomplete statements below is followed by five possible answers or completions. For both Biology-E and Biology-M, select the one choice that is the best answer and fill in the corresponding space on the answer sheet.

1. Referring to this list of vertebrates, which is the correct sequence of evolution?

    (A) Bony fish—amphibians—reptiles—birds
    (B) Birds—bony fish—amphibians—reptiles
    (C) Amphibians—reptiles—bony fish—birds
    (D) Reptiles—birds—bony fish—amphibians
    (E) Reptiles—birds—amphibians—bony fish

2. Which of the following is a homeotherm?

    (A) Grasshopper
    (B) Hydra
    (C) Earthworm
    (D) Blue jay
    (E) Frog

3. Mycorrhizae are

    (A) plants that have no vascular tissue
    (B) nitrogen-fixing bacteria that live in nodules on the roots of legumes
    (C) primitive plants like mosses, which show a dominant gametophyte stage
    (D) vascular bundles in the stems of tracheophytes
    (E) symbiotic structures living in the roots of plants that increase uptake of nutrients from the soil

4. All of the following are matched correctly EXCEPT

   (A) nerve net—hydra
   (B) Malpighian tubules—earthworms
   (C) nematocysts—hydra
   (D) contractile vacuoles—amoeba
   (E) flame cells—planaria

5. The tissue in a plant that constantly undergoes mitosis is the

   (A) pith
   (B) xylem
   (C) phloem
   (D) cortex
   (E) cambium

6. According to Hardy-Weinberg theory, which of the following represents a heterozygous individual?
   (A) p
   (B) p^2
   (C) $2pq$
   (D) q^2
   (E) q

7. In the case of pea plants, tall (T) is dominant over dwarf (t). What is the genotype of the parents of a generation of plants half of which are tall and half of which are dwarf?

   (A) $Tt \times tt$
   (B) $Tt \times Tt$
   (C) $TT \times tt$
   (D) $X^T X^t \times X^T X^t$
   (E) $X^T X^T \times X^t X^t$

8. Within less than 2 years of the introduction of a new antibiotic, bacteria appear that are resistant to that antibiotic. This is an example of

   I. Divergent evolution
   II. Adaptive radiation
   III. Directional selection

   (A) I only
   (B) II only
   (C) III only
   (D) I and III only
   (E) I, II, and III

9. For which of the following pairs is the first term NOT a building block of the second term?

   (A) Fatty acid—insulin
   (B) Glucose—chitin
   (C) Thymine—nucleotide
   (D) Amino acid—hemoglobin
   (E) Nitrogen—uric acid

10. A boy with red-green color blindness has a color-blind father and a mother who is not color-blind. The boy inherited his color blindness from

    (A) his father
    (B) his mother
    (C) either his father or his mother
    (D) both parents; this is an example of incomplete dominance
    (E) it cannot be determined

11. All of the following are related to locomotion EXCEPT

    (A) setae
    (B) tendons
    (C) pseudopods
    (D) hydrostatic skeleton
    (E) typhlosol

12. During which phase of the cell cycle does DNA replication occur?

    (A) Prophase
    (B) Metaphase
    (C) Cytokinesis
    (D) Interphase
    (E) Telophase

Directions: Each set of lettered choices below refers to the numbered questions or statements immediately following it. Select the one lettered choice that best answers each question and fill in the corresponding space on the answer sheet. A choice may be used once, more than once, or not at all in each set.

Questions 13–16

   (A) Global warming
   (B) Eutrophication of lakes
   (C) Depletion of the ozone layer
   (D) Magnification in the food chain
   (E) Acid rain

13. Related to the use of chlorofluorocarbons

14. Pesticides sprayed on crops can cause illness in people

15. Caused by an increase in CO_2 concentration in the atmosphere

16. Caused by SO_2 in the air

Questions 17–18

    I. Acoelomate
   II. Radial symmetry
 III. Cephalization

17. Cnidarians are characterized by which of the following?

    (A) I only
    (B) II only
    (C) III only
    (D) I and III only
    (E) I, II, and III

18. Flatworms are characterized by which of the following?

    (A) I only
    (B) II only
    (C) III only
    (D) I and III only
    (E) I, II, and III

Questions 19–21

    (A) Ethylene gas
    (B) Abscisic acid
    (C) Auxin
    (D) Gibberellins
    (E) Cytokinins

19. Enhances apical dominance

20. The reason that "One bad apple spoils the whole barrel."

21. Responsible for phototropisms

Questions 22–24

    (A) Transformation
    (B) Translation
    (C) Transcription
    (D) Translocation

22. Occurs at the ribosome in eukaryotes

23. DNA codes for mRNA

24. Ability of bacteria to uptake genes from other cells

Questions 25–26

An experiment is carried out to explore inheritance of the bar-eyed trait in fruit flies. Homozygous bar-eyed and wild-type flies are mated, producing two F_1 generations. Then those F_1 offspring are mated with each other, producing two F_2 generations. One hundred offspring from each cross are recorded. Here is the data from those 4 crosses.

Cross 1	Cross 2
Parents **Bar-eyed Female × Wild-type Male**	**Parents** **Wild-type Female × Bar-eyed Male**
F_1 53 Bar-eyed males 47 Bar-eyed females	F_1 48 Wild-type males 52 Bar-eyed females
Cross 3 **F_1 flies from cross 1 were mated**	**Cross 4** **F_1 flies from cross 2 were mated**
F_2 22 Bar-eyed males 51 Bar-eyed females 27 Wild-type males 0 Wild-type females	F_2 25 Bar-eyed males 26 Bar-eyed females 27 Wild-type males 25 Wild-type females

25. The pattern of inheritance for bar-eyed is

    (A) autosomal dominant
    (B) autosomal recessive
    (C) sex-linked dominant
    (D) sex-linked recessive
    (E) cannot be determined by the information given

26. What is the most likely genotype for a bar-eyed female (underlined) in the F_1 generation in cross 1?

    (A) BB
    (B) Bb
    (C) Wb
    (D) $X^B X^b$
    (E) $X^b X^b$

Questions 27–28

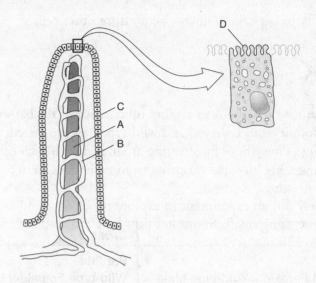

27. Which of the following is true about the structure shown in the figure?

    (A) It is found in the small intestine.
    (B) It is part of the large intestine.
    (C) It is necessary for protein digestion.
    (D) It is part of the hepatic system.
    (E) It is a flame cell.

28. Which is true about the structure at *A*?

    (A) It absorbs sugars.
    (B) It is a lacteal.
    (C) It hydrolyzes vitamins.
    (D) It is a capillary.
    (E) It releases hydrolytic enzymes.

Questions 29–30

    (A) Pancreas
    (B) Adrenal medulla
    (C) Anterior pituitary
    (D) Posterior pituitary
    (E) Hypothalamus

29. Regulates blood sugar levels

30. Bridge between nervous and endocrine systems

**If you are taking the Biology-E test, continue with questions 31–40.
If you are taking the Biology-M test, go to question 41 now.**

Biology-E Section

31. All of the following are characteristics of populations EXCEPT

    (A) size
    (B) density
    (C) age distribution
    (D) phenotype
    (E) death rate

Questions 32–34

Students carry out an experiment to explore the effect of different environments on blood flow using small, freshwater fish. Each fish is placed into a petri dish. A wet cotton ball is placed over its gills to allow for the diffusion of oxygen, which will keep the fish alive. The students then place the petri dish onto the stage of a light microscope so that the thinnest part of the tail is directly under the objective lens and blood can be seen flowing in blood vessels. While focusing on a capillary, students count the number of red blood cells flowing through the vessel at 5-second intervals. After monitoring the normal flow of blood for 30 seconds, the environment in the petri dish is altered in one of five different ways. Students observe and record the change in blood flow, if any. After the experiment, each fish is returned to the fish tank.

Experiment 1—Nothing is added to this petri dish.
Experiment 2—Water at a temperature of 5°C is added.
Experiment 3—Water at a temperature of 30°C is added.
Experiment 4—Ethyl alcohol is added.
Experiment 5—Nicotine extracted from cigarette tobacco is added.

Experiment	Initial Average Rate of Blood Flow	Final Average Rate of Blood Flow
1	9	9
2	8	4
3	9	13
4	8	5
5	9	12

32. Which graph below best describes what happened in experiment 4?

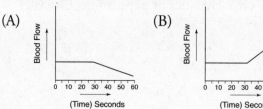
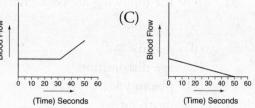
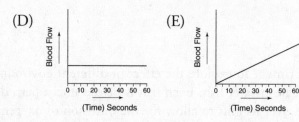

33. All of the following statements about this experiment are correct EXCEPT

    (A) alcohol is a central nervous system depressant and decreases the blood flow in the capillary
    (B) nicotine is a central nervous system stimulant and increases blood flow in the capillary
    (C) increasing the temperature increases the blood flow in the capillary
    (D) the changes in blood flow in experiments 2 and 3 are temporary because a fish can adjust its body temperature
    (E) experiment 1 is the control

34. The dependent variable in experiment 3 is

    (A) the health of the individual fish
    (B) temperature of the water
    (C) nicotine
    (D) alcohol
    (E) rate of blood flow

Questions 35–37

    (A) Stabilizing selection
    (B) Directional selection
    (C) Disruptive selection
    (D) Genetic drift
    (E) Convergent evolution

35. The majority of human birth weights is between 6 and 9 pounds

36. The peppered moths in England in the 20th century

37. Founder and bottleneck effect are examples

Questions 38–40

Epidemiologists concerned with the spread of the hantavirus from rats to humans conducted an experiment to explore the effectiveness of a particular pesticide. They exposed a population of 100 rats to the pesticide on day 1. After exposure, they allowed any rats that survived to reproduce. They monitored the population of rats for 100 days and plotted a graph of the data they collected.

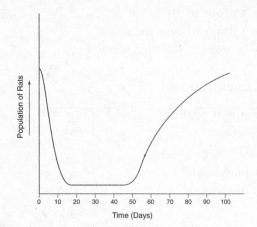

38. The data suggest that on day 1

    (A) the female rats were sterilized by the pesticide
    (B) the male rats were sterilized by the pesticide
    (C) most of the rats were sterilized by the pesticide
    (D) all the rats were killed by the pesticide
    (E) most of the rats were killed by the pesticide

39. The best explanation for the results of this experiment is

    (A) some rats developed a resistance to the pesticide
    (B) some rats were resistant to the pesticide at the outset of the experiment
    (C) no rats were resistant to the pesticide
    (D) none of the rats were able to reproduce
    (E) the rats evolved a resistance because they needed to

40. Which of the following biological processes is illustrated by this experiment?

    (A) Biological magnification
    (B) Theory of use and disuse
    (C) Natural selection
    (D) Ecological succession
    (E) Punctuated equilibrium

This is the end of the mini-test. Check over your work on the entire test.
Grade your test now.

Biology-M Section

<u>Questions 41–44</u>

Wet mounts of three living samples of elodea cells are prepared for viewing under the light microscope. Each slide is mounted with a different solution and viewed after 5 minutes.

Sample A—Elodea + 5 drops of 10% NaCl
Sample B—Elodea + 5 drops of isotonic saline
Sample C—Elodea + 5 drops of distilled water

41. The cells in sample A would

    (A) exhibit turgor pressure
    (B) undergo lysis
    (C) swell and burst
    (D) exhibit plasmolysis
    (E) remain unchanged

42. The results of the experiment illustrate which of the following processes?

    (A) Hydrolysis
    (B) Dehydration
    (C) Active transport
    (D) Polymerization
    (E) Osmosis

43. Which cell structure(s) would be visible in sample A that was not visible prior to exposure to the 10% NaCl?

    (A) Nucleus
    (B) Plasma membrane
    (C) Mitochondria
    (D) Golgi body
    (E) Chloroplasts

44. Which of the following statements about this experiment is correct?

    (A) The movement of salt is the dependent variable
    (B) The elodea cells in sample C are in a hypotonic solution
    (C) There is no passage of water in either direction in sample B
    (D) NaCl is a toxin that would destroy any cell, no matter the concentration
    (E) You cannot predict with any certainty what would happen to these cells; they are living organisms

45. Which of the following statements best explains the fact that a mutation in a cell's DNA does not always result in an error in the polypeptide produced from that DNA sequence?

    (A) Some polypeptides are produced by a code other than a nucleic acid code.
    (B) The nucleolus can repair damaged DNA.
    (C) The Golgi body can repair damaged DNA.
    (D) Different codons code for more than one amino acid.
    (E) Scientists have no idea why this phenomenon occurs.

46. All of the following disorders are caused by a mutation in the DNA sequence EXCEPT

    (A) sickle cell anemia
    (B) PKU
    (C) cystic fibrosis
    (D) AIDS
    (E) hemophilia

47. The DNA sequence is converted into an amino acid sequence in eukaryotic cells at the

    (A) ribosome
    (B) cytoplasm
    (C) nucleus
    (D) endoplasmic reticulum
    (E) peroxisome

Questions 48–50

    I. Glycolysis
    II. Krebs cycle
    III. Electron transport chain

48. ATP is produced by oxidative phosphorylation.

    (A) I only
    (B) II only
    (C) III only
    (D) II and III only
    (E) I, II, and III

49. The ATP synthetase channel produces ATP.

    (A) I only
    (B) II only
    (C) III only
    (D) II and III only
    (E) I, II, and III

50. Takes place in mitochondria

    (A) I only
    (B) II only
    (C) III only
    (D) II and III only
    (E) I, II, and III

STOP

This is the end of the mini-test. Check over your work on the entire test. **Grade your test now.**

GRADE YOUR PRACTICE EXAM

You can earn anywhere from 200 to 800 points on the SAT Subject Test: Biology E/M. To figure out what you scored on your mini-diagnostic test, you first need to determine your raw score.

Determine Your Raw Score

Step 1: Compare your answer sheet to the correct answers on Table 3.1.

- Put a check in the column marked "Right" if your answer is correct.
- Put a check in the column marked "Wrong" if your answer is incorrect.
- Leave both columns blank if you omitted the question.

Step 2: Count the number of right answers, and enter the number here:　　　　　　　　　　　　　　　　　　_____

Step 3: Count the number of wrong answers, divide by 4, and enter the number here:　　　　　　　　　　　　　　　_____

Step 4: Subtract the number you obtained in Step 3 from the number in Step 2. Round the result to the nearest whole number (0.5 is rounded up), and enter here:　　　　　_____

Step 5: The number you obtained in Step 4 is your raw score. Convert it to your College Board Score using Table 3.2, which is similar to those published by the College Board.

TABLE 3.1

Question No.	Correct Answer	Right	Wrong	Question No.	Correct Answer	Right	Wrong
The Correct Answers							
1	A			26	D		
2	D			27	A		
3	E			28	B		
4	B			29	A		
5	E			30	E		
6	C			31	D		
7	A			32	A		
8	C			33	D		
9	A			34	E		
10	B			35	A		
11	E			36	B		
12	D			37	D		
13	C			38	E		
14	D			39	B		
15	A			40	C		
16	E			41	D		
17	B			42	E		
18	D			43	B		
19	C			44	B		
20	A			45	D		
21	C			46	D		
22	B			47	A		
23	C			48	C		
24	A			49	C		
25	C			50	D		

TABLE 3.2

Score Conversion Table

Raw Score	Scaled Score	Raw Score	Scaled Score	Raw Score	Scaled Score
50	800	30	620	10	420
49	800	29	610	9	410
48	790	28	600	8	400
47	790	27	590	7	390
46	780	26	580	6	380
45	770	25	570	5	370
44	760	24	560	4	360
43	750	23	550	3	350
42	740	22	540	2	340
41	730	21	530	1	330
40	720	20	520	0	320
39	710	19	510	−1	310
38	700	18	500	−2	300
37	690	17	490	−3	290
36	680	16	480	−4	280
35	670	15	470	−5	270
34	660	14	460	−6	260
33	650	13	450	−7	250
32	640	12	440	−8	240
31	630	11	430	−9	200

EXPLANATION OF ANSWERS

1. **(A)** Organisms first evolved in water and moved to land much later. Bony fish are the most primitive. Amphibians live part of their lives on land and part in water; they must return to a watery environment to reproduce. Reptiles evolved legs, scales, and the shelled egg that enabled them to move to land permanently. Birds evolved from reptiles and took to the skies to avoid competition with land dwellers

2. **(D)** Homeotherm means warm-blooded. It refers to an animal that regulates its body temperature. Birds and mammals are homeotherms. The others are not.

3. **(E)** Mycorrhizae are symbionts that live in the roots of most plants and help in the uptake of nutrients. Plants that have no vascular tissue are called bryophytes. Nitrogen-fixing bacteria are another type of symbiont that live in the roots of a group of plants known as legumes.

4. **(B)** Malpighian tubules are structures found in insects, like grasshoppers, that aid in excretion. They excrete nitrogenous wastes into the digestive tract and from there, out of the body. All the other pairs are correctly matched. Nematocysts are stingers, and the flame cell is a structure for excretion in flatworms like planaria.

5. **(E)** The cambium is the growth tissue in plants located between the xylem and phloem. It is always dividing. Cambium is a specific type of meristematic tissue.

6. **(C)** Let's use the trait for height in plants to explain this. T is dominant and represents tall; t is recessive and represents dwarf. The letter p represents the dominant trait (T), so p^2 represents homozygous dominant (TT). The letter q represents the recessive trait (t), so q^2 represents homozygous recessive (tt). The hybrid condition is represented by $2pq$. This concept is based on the monohybrid cross. To satisfy the conditions in the question, two squares must contain Tt, which equals $2pq$ (Tt).

	T	t
T	TT	Tt
t	Tt	tt

7. **(A)** Work this out by doing a Punnett square in reverse. First, set up the square and fill in whatever you can about the offspring given what is described in the question.

T	tt
T	tt

Then divide out to the parental genotypes on the outside of the Punnett square.

	T	t
t	Tt	tt
t	Tt	tt

The trait is not sex-linked, so you do not use the X and Y notation.

8. **(C)** The population of bacteria changed over time and became resistant to the antibiotic. This is an example of directional selection. Another example of this is the peppered moths in England from 1835 to 1900. The moth population rapidly changed from light to dark because of a rapid change in the environment.

9. **(A)** Fatty acids are building blocks of lipids, fats, and waxes. Insulin is a protein. The building blocks of it are amino acids.

10. **(B)** Boys inherit the Y chromosome from their fathers and the X chromosome from their mothers. A father cannot give his son a sex-linked condition. The son can inherit a sex-linked condition only from his mother. Color blindness is sex-linked. In this case, although she did not have the condition, his mother must have been a carrier.

11. **(E)** The typhlosol is a structure in the intestine of the earthworm that greatly enhances absorption of nutrients into the bloodstream. Setae are hairlike structures in each segment of the earthworm that help push the animal along. The hydrostatic skeleton helps soft-bodied animals, like worms, move. The body cavity fills with fluid and makes the animal more rigid. Pseudopods are characteristics of amoeba. Tendons connect skeletal muscle to bone.

12. **(D)** Interphase is sometimes referred to as the "resting stage." However, the cell is certainly not at rest. It is actively carrying out all the life functions and also replicating DNA in preparation for cell division. Most of the life of a cell is spent in interphase.

13. **(C)** Chlorofluorocarbons are now banned. They cause the depletion of the ozone layer.

14. **(D)** Chemicals sprayed on crops get incorporated into the food chains and accumulate in animals that feed at the top of the chain.

15. **(A)** Earth's climate has changed, becoming warmer, as a result of the greenhouse effect, which results from the accumulation of CO_2 in the atmosphere.

16. **(E)** SO_2 mixes with water vapor in the air and produces sulfurous acid and sulfuric acid that rain down as acid rain.

17. **(B)** Cnidarians are very primitive animals and exhibit radial symmetry. They are only two cell layers thick and do not undergo embryonic development like flatworms and more advanced animals. The flatworms are three layers thick and are acoelomate.

18. **(D)** Flatworms are acoelomates. They do not have a coelom; but they do exhibit cephalization.

19. **(C)** Auxins are growth hormones. They are responsible for tropisms and apical dominance.

20. **(A)** Ethylene gas is the plant hormone that promotes fruit ripening and rotting.

21. **(C)** Auxins are growth hormones. Tropisms are caused by an unequal distribution of auxins.

22. **(B)** Translation is the process by which codons of an mRNA sequence are changed into an amino acid sequence. This occurs at the ribosomes.

23. **(C)** Transcription is the process by which DNA makes RNA. This occurs in the nucleus.

24. **(A)** Transformation is a natural process whereby bacteria uptake genes from other bacteria. This was discovered by Griffith in 1927 when he was studying the bacterium that causes pneumonia.

25. **(C)** Crosses 1 and 2 are called reciprocal because the phenotypes of the male and female are the opposite in each cross. The first thing you should notice about the offspring (F_1) of crosses 1 and 2 is that the results are different from each other. In addition, the phenotypes of male and female offspring are different from each other. These two facts tell you that the inheritance of the bar-eyed trait is probably sex-linked. Next, look at cross 2. The females (F_1) inherited the bar-eyed trait from their male parent and they express the bar-eyed trait. They are not merely carriers. Their genotypes must be X–X or X^BX^b. Therefore, the bar-eyed trait must be dominant, sex-linked dominant. Here are the Punnett squares for all the crosses.

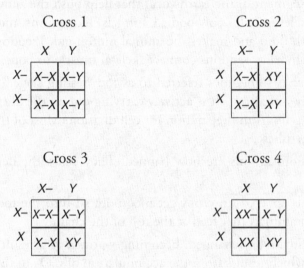

26. **(D)** See the explanation for #25.

27. **(A)** This structure is a villus. Millions of them line the small intestine and give the lining of the small intestine a fuzzy appearance, called a brush border. They greatly enhance absorption into the bloodstream by increasing the surface area.

28. **(B)** Structure *A* is a lacteal. It is a part of the lymphatic system. The lacteal absorbs fatty acids and glycerol. The capillary in the villus absorbs monosaccharides and amino acids. See the explanation for #27.

29. **(A)** The pancreas contains the islets of Langerhans. These contain special cells that release either glucagon (which raises blood sugar) or insulin (which lowers blood sugar).

30. **(E)** The hypothalamus is the bridge between the nervous and endocrine systems. As part of the endocrine system, it produces and releases the hormones oxytocin and ADH that are stored in the posterior pituitary until needed. As part of the nervous system, the hypothalamus electrically stimulates the anterior pituitary to release the many hormones that gland produces.

31. **(D)** Phenotype means how an organism appears. It is a property of an individual, not a population. The others are characteristics of populations.

32. **(A)** In experiment 4, blood flow decreases after 30 seconds. Graph C shows a decrease beginning at time 0.

33. **(D)** The fish is cold-blooded and cannot change its body temperature. All the other statements are correct.

34. **(E)** The dependent variable is the one that changes in response to changes in the experiment.

35. **(A)** Stabilizing selection eliminates the extremes and favors the more common intermediate forms. Many mutants are weeded out.

36. **(B)** One phenotype replaces another in the gene pool. The peppered moths are a famous example. Another example is a population of bacteria that becomes resistant to a particular antibiotic.

37. **(D)** Genetic drift is a change in a gene pool due to chance. Two examples are the bottleneck effect and the founder effect.

38. **(E)** Most, but not all of the rats, were killed by the pesticide. You have no evidence that any rats were sterilized because they continue to reproduce and the population increases in size again.

39. **(B)** Some rats apparently were resistant to the pesticide. They had the selective advantage after exposure to the pesticide; those that were susceptible to the pesticide died. The only rats that survived to pass their genes for resistance to their offspring were the ones that were resistant.

40. **(C)** After exposure to the pesticide, those that were susceptible to the pesticide died. The only rats that survived to pass their genes for resistance to their offspring were the ones that were resistant. This is a perfect example of natural selection.

41. **(D)** Plasmolysis is cell shrinking due to loss of water from a cell. Water leaves the cell because the elodea is in a hypertonic environment with lower concentration of water outside the cell than inside the cell.

42. **(E)** Osmosis is specifically the diffusion of water. Water always flows passively, down a gradient, from a high concentration of water to a low concentration of water.

43. **(B)** Once water leaves the elodea cell, there is nothing to keep the plasma membrane against the cell wall. The membrane collapses into the center of the cell with the other remaining organelles. At this point, the plasma membrane is clearly visible.

44. **(B)** The elodea in sample C is in distilled water, a hypotonic environment. Water will flow into the cells because there is more water/less solute in the surrounding solution.

45. **(D)** There are 64 different codons but only 20 different amino acids. So several codons code for the same amino acid. For example, ACC, ACU, ACG, and ACA all code for the same amino acid, threonine. A point mutation in the DNA, such as a change from AGG to AGA, would still translate to the same amino acid, threonine.

46. **(D)** AIDS is caused by the HIV virus. All the other diseases are inherited and are caused by mutations in the gene or genes that code for a particular enzyme.

47. **(A)** Translation is the process by which the DNA code carried by mRNA from the nucleus is changed into an amino acid sequence, a polypeptide, at the ribosome.

48. **(C)** ATP is produced by the process known as oxidative phosphorylation in the electron transport chain. During glycolysis and the Krebs cycle, ATP is produced by a process known as substrate-level phosphorylation. Oxidative phosphorylation produces ATP using the ATP-synthetase structure in both the cristae membrane of mitochondria and the thylakoid membrane in chloroplasts.

49. **(C)** ATP is produced by the process known as oxidative phosphorylation in the electron transport chain. It relies on the ATP synthetase channel in the cristae membrane of mitochondria and thylakoid membranes in chloroplasts.

50. **(D)** The Krebs cycle occurs in the inner matrix of the mitochondria, and the electron transport chain occurs within the cristae membrane and outer compartment of the same organelle. Glycolysis occurs in the cytoplasm.

What Topics Do You Need to Work On?

Table 3.3 shows an analysis by topic for each question on the test you just took.

TABLE 3.3

Topic Analysis

Cellular and Molecular Biology	Heredity	Evolution and Diversity	Organismal Biology	Ecology
9, 12, 41, 42, 43, 44, 48, 49, 50	7, 10, 22, 23, 24, 25, 26, 45, 46, 47	1, 2, 6, 8, 17, 18, 35, 36, 37	3, 4, 5, 11 19, 20, 21, 27, 28, 29, 30, 32, 33, 34	13, 14, 15, 16, 31, 38, 39, 40

CELLULAR AND MOLECULAR BIOLOGY

Biochemistry

CHAPTER 4

- Basic atomic structure
- Bonding
- Characteristics of water
- pH
- Carbohydrates
- Lipids
- Proteins and enzymes
- Nucleic acids

To understand about living things, you must understand some basic biochemistry. Biochemistry affects every aspect of our lives every day. Sweating cools the skin because of strong hydrogen bonding between water molecules. Your body maintains your blood at one critical pH because of the bicarbonate buffering system. To prevent heart attacks, you must begin with an understanding about the structure of fatty acids. Mad cow disease is caused by a misfolded protein. Here is a review of basic biochemistry.

BASIC CHEMISTRY—ATOMIC STRUCTURE

The atom consists of subatomic particles: protons, neutrons, and electrons. Table 4.1 compares these particles.

TABLE 4.1

Subatomic Particles			
Subatomic Particle	Charge	Mass in amu	Location
Proton	+1	1	Nucleus
Neutron	0	1	Nucleus
Electron	−1	0	Outside nucleus

1. An atom in the elemental state always has a neutral charge because the number of protons (+) equals the number of electrons (−).

2. Electron configuration is important because it determines how a particular atom will react with atoms of other elements.

3. Electrons in the lowest available energy level are said to be in the **ground state**.

4. When an atom absorbs energy, its electrons move to a higher energy level,. The atom is then said to be in the **excited state**. For example, during photosynthesis, chlorophyll molecules absorb light energy, which boosts electrons to higher energy levels. These excited electrons provide the energy to make sugar as they return to their ground state and release the energy they previously absorbed.

Isotopes are atoms of one element that vary only in the number of neutrons in the nucleus. Chemically, all isotopes of the same element are identical because they have the same number of electrons. For example, carbon-12 and carbon-14 are isotopes of each other and are chemically identical. They both possess 6 protons and 6 electrons. However, carbon-12 has 6 neutrons, while carbon-14 has 8 neutrons. Some isotopes, like carbon-14, are radioactive (called **radioisotopes**). The nuclei of radioisotopes emit particles and decay at a known rate called a **half-life**. Knowing the half-life enables us to measure the age of fossils or to estimate the age of Earth.

Radioisotopes are useful in other ways. Besides measuring the age of fossils, they can be used in medical diagnosis, treatment, and research. For example: radioactive iodine (I-131) can be used both to diagnose and to treat certain diseases of the thyroid gland. In addition, radioactive carbon can be used as a **tracer**, incorporated into molecules of carbon dioxide, and used to track metabolic pathways.

> **CAREFUL**
>
> Don't confuse isotope with isomer (see page 42).

BONDING

A bond is formed when two atomic nuclei attract the same electron(s). Energy is released when a bond is formed. Energy must be supplied or absorbed to break a bond. Atoms bond to achieve stability, to acquire a completed outer shell. There are two main types of bonds, ionic and covalent.

Ionic bonds form when electrons are transferred. An atom that gains electrons becomes an **anion**, which stands for a negative ion. An atom that loses an electron becomes a **cation**, a positive ion. Ions, such as Cl^-, Na^+, and Ca^{2+} are necessary for normal cell, tissue, and organ function.

Covalent bonds form when atoms share electrons. The resulting structure is called a **molecule**. A single covalent bond (–) results when two atoms share one pair of electrons. A double covalent bond (=) results when two atoms share two pairs of electrons, and a triple covalent bond (≡) results when two atoms share three pairs of electrons.

Table 4.2 describes two types of covalent bonds.

TABLE 4.2

The Two Types of Covalent Bonds	
Nonpolar Covalent Bond	**Polar Covalent Bond**
Electrons shared equally	Electrons shared unequally
Formed between any two atoms that are alike	Formed between any two atoms that are unalike
Examples:	Examples:
H_2 (H – H) and O_2 (O = O)	CO (C = O) and H_2O (H – O – H)
nonpolar bond ↑ ↑ nonpolar bond	polar bond ↑ ↑ ↑ polar bonds

Intermolecular Attractions

Not only do atoms within molecules attract each other. Individual molecules attract each other in a variety of ways via **intermolecular attractions**.

POLAR-POLAR ATTRACTION

When two or more atoms form a bond, the entire resulting molecule is either polar (unbalanced) or nonpolar (balanced). Strong attractions exist between polar molecules. The negative end of one polar molecule attracts the positive end of another polar molecule. H_2O is a highly polar (unbalanced) molecule. It looks like this:

$$\overset{-}{O}$$
$$\underset{+}{H} \qquad \underset{+}{H}$$

HYDROGEN BONDING

Hydrogen bonding is very important to living things. It occurs between certain molecules containing atoms that exert a strong pull on their atoms.

Hydrogen bonding:

- Keeps the two strands of DNA bonded together, forming a double helix.
- Causes water molecules to stick together and is responsible for many special characteristics about water as shown in Figure 4.1.

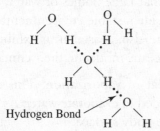

Figure 4.1 Hydrogen bonding

NONPOLAR MOLECULES

Only the weakest attractions (van der Waals) exist between nonpolar molecules. An example of a nonpolar molecule is CO_2. It is linear and balanced. It looks like this:

$$O = C = O$$

HYDROPHOBIC AND HYDROPHILIC

Hydrophobic means "water hating" or "repelled by water." **Hydrophilic** means "water loving" or "attracted to water." Substances that are polar will dissolve in water, while substances that are nonpolar will not dissolve in water. Like dissolves like. You are very familiar with one example. Carbon dioxide (nonpolar), the molecule that gives soda pop its fizziness, does not dissolve in water (polar), which is why gas escapes when you open a can of soda pop and it goes flat.

Lipids, which are nonpolar, are hydrophobic and do not dissolve in water, which is why oil and vinegar salad dressing separates upon standing. Since the plasma membrane is a phosholipid bilayer, only nonpolar substances can readily dissolve through the plasma membrane. Large polar molecules must travel across a membrane in special hydrophilic (protein) channels.

CHARACTERISTICS OF WATER

Water is asymmetrical and very polar. It also has strong intermolecular attractions. In addition to polar attractions, water exhibits strong hydrogen bonding, as shown in Figures 4.1 and 4.2. Together, these two forces are responsible for the special characteristics of water that affect life on Earth.

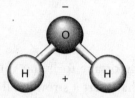

Figure 4.2 Water molecule

1. **Water has a high specific heat.** Specific heat is the amount of heat that must be absorbed in order for 1 gram of a substance to change its temperature 1° Celsius. This means that large bodies of water, like oceans, resist changes in temperature and provide a stable environmental temperature for the organisms that live in them. Also, coastal areas exhibit relatively little temperature change because the oceans moderate their climates.

2. **Water has a high heat of vaporization.** This means that a relatively great amount of heat is needed to evaporate water. As a result, evaporation of sweat significantly cools the body surface.

3. **Water has high adhesion properties.** Adhesion is the clinging of one substance to another, and it plays an important role in plant survival. Forces of adhesion contribute to capillary action, which helps water flow up from the roots of a plant to the leaves.

4. **Water is the universal solvent.** Because water is a highly polar molecule, it dissolves all polar and ionic substances.

5. **Water exhibits strong cohesion tension.** This means that molecules of water tend to stick together. This results in several biological phenomena. Water moves up a tall tree from the roots to the leaves without the expenditure of energy by what is referred to as transpirational-pull cohesion tension. It also results in surface tension that allows insects to walk on water without breaking the surface.

6. **Ice floats because it is less dense than water.** In a deep body of water, floating ice insulates the liquid water below it, allowing life to exist beneath the frozen surface during cold seasons. The fact that ice covers the surface of water in a lake in the cold months and melts in the spring results in a stratification of the lake during the winter and considerable mixing in the spring. In the spring, surface ice melts, becomes denser water, and sinks to the bottom of the lake, causing water to circulate throughout. Oxygen from the surface is returned to the depths, and nutrients released by the activities of bottom-dwelling bacteria are carried to the upper layers of the lake. This cycling of the nutrients in the lake is known as the spring overturn and is necessary to the health of a lake.

pH

pH is a measure of acidity and alkalinity of a solution. It is the negative logarithm of the hydrogen ion concentration in moles per liter. Anything with a pH of less than 7 is acidic, and anything with a pH value greater than 7 is alkaline or basic. A substance with a pH of 7 is neutral; see Figure 4.3.

> **REMEMBER**
>
> As the concentration of H^+ increases, the pH decreases.

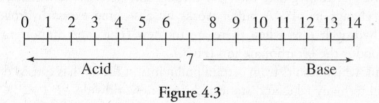

Figure 4.3

As shown in Table 4.3, a solution of pH 1 is 10 times more acidic than a solution with a pH of 2, and 100 times more acidic than a solution with a pH of 3. It is 1,000 times more acidic than a solution with a pH of 4.

TABLE 4.3

pH Compared with Molarity	
pH	**Concentration of H⁺ ions in Moles per Liter**
1	$1 \times 10^{-1} = 0.1$ molar
2	$1 \times 10^{-2} = 0.01$ molar
3	$1 \times 10^{-3} = 0.001$ molar
4	$1 \times 10^{-4} = 0.000\,1$ molar
7	$1 \times 10^{-7} = 0.000\,000\,1$ molar
13	$1 \times 10^{-13} = 0.000\,000\,000\,000\,1$ molar

The pH of some common substances is shown in Table 4.4.

TABLE 4.4

pH Values	
Substance	pH
Stomach acid	2
Orange juice	3.5
Carbonated drinks	3.0
Acid rain	<5.6
Milk	6.5
Seawater	8.5
Human blood	7.4

The internal pH of most living cells is close to 7. Even a slight change can be harmful.

Biological systems regulate their pH through the presence of **buffers**, substances that resist change in pH. A buffer works by absorbing excess hydrogen ions or donating hydrogen ions when there are too few. The most important buffer in human blood is the **bicarbonate ion** (HCO_3^-).

Acid rain, which results from certain pollutants in the air, has caused damage and destruction to many lakes and stone architecture worldwide.

ORGANIC COMPOUNDS

Organic compounds are compounds that contain carbon. There are four classes of organic compounds: carbohydrates, lipids, proteins, and nucleic acids.

Carbohydrates

Carbohydrates consist of only three elements: carbon, hydrogen, and oxygen.

- Carbohydrates supply quick energy.
- One gram of any carbohydrate will release 4 calories when burned.
- Dietary sources include rice, pasta, bread, and cookies.
- There are three classes of carbohydrates: monosaccharides, disaccharides, and polysaccharides.

MONOSACCHARIDES

All monosaccharides have a chemical formula of $C_6H_{12}O_6$. Examples are glucose, galactose, and fructose, which are **isomers** of each other. Isomers are compounds with the same molecular formula, but with different structures. Therefore, they have different physical and chemical properties.

The structural formula of glucose is shown in Figure 4.4.

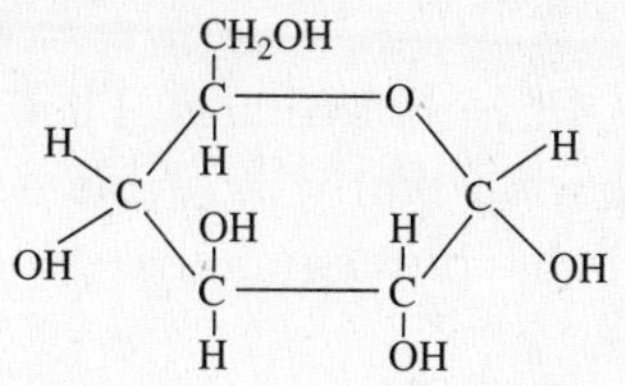

Figure 4.4 Monosaccharide

DISACCHARIDES

All disaccharides have the chemical formula $C_{12}H_{22}O_{11}$. They consist of two monosaccharides joined by a process known as **dehydration synthesis.**

The structure of the disaccharide, maltose, is shown in Figure 4.5.

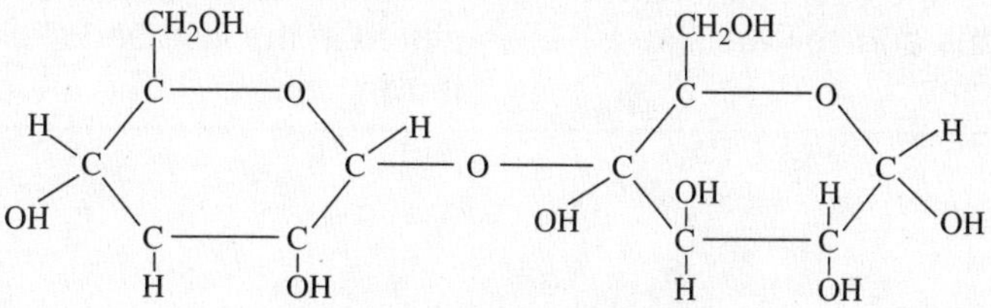

Figure 4.5 Disaccharide

Table 4.5 shows three dehydration synthesis reactions of monosaccharides. They produce three disaccharides: maltose, lactose, and sucrose.

TABLE 4.5

Dehydration Synthesis of Monosaccharides				
Monosaccharides + Monosaccharide → Disaccharide + Water				
$C_6H_{12}O_6$	+	$C_6H_{12}O_6$	→	$C_{12}H_{22}O_{11}$	+	H_2O
Glucose	+	Glucose	→	Maltose	+	Water
Glucose	+	Galactose	→	Lactose	+	Water
Glucose	+	Fructose	→	Sucrose	+	Water

Hydrolysis is the breakdown of a compound. It is what occurs during **digestion** and is the reverse of dehydration synthesis.

Sucrose + Water → Glucose + Fructose

POLYSACCHARIDES

Polysaccharides are polymers of carbohydrates. They form as many monosaccharides join together by dehydration synthesis. Table 4.6 shows the four important polysaccharides: cellulose, starch, chitin, and glycogen.

TABLE 4.6

Polysaccharides

Found in Plants	Cellulose	Starch
	Makes up plant cell walls	The way sugar is stored in plants
Found in Animals	**Chitin**	**Glycogen**
	Makes up the exoskeleton in arthropods and cell walls in mushrooms	"Animal starch"; in humans, this is stored in the liver and skeletal muscle

Lipids

Lipids are a diverse class of organic compounds that include fats, oils, and waxes. Structurally, most lipids consist of one glycerol and three fatty acids, as shown in Figure 4.6.

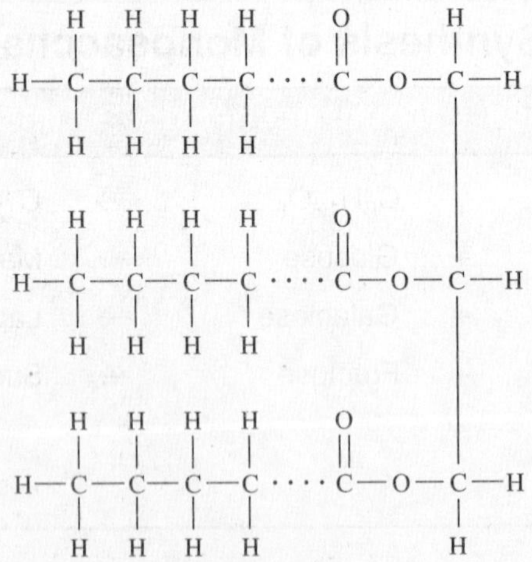

Figure 4.6

Glycerol is an alcohol whose structure is shown in Figure 4.7.

$$
\begin{array}{c}
H \\
| \\
H\!-\!C\!-\!OH \\
| \\
H\!-\!C\!-\!OH \\
| \\
H\!-\!C\!-\!OH \\
| \\
H
\end{array}
$$

Figure 4.7 Glycerol

A fatty acid is a hydrocarbon chain with a carboxyl group at one end. Fatty acids exist in two varieties: saturated and unsaturated. These are shown in Figure 4.8.

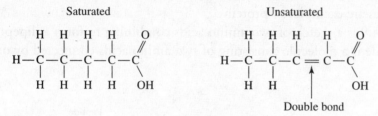

Figure 4.8 Fatty acids

SATURATED FATS

Saturated fats, with a few exceptions, come from animals. They are solid at room temperature and when ingested in large quantities, are linked to heart disease. An example of a saturated fat is butter. Saturated fatty acids contain only single bonds between carbon atoms.

UNSATURATED FATS

Unsaturated fatty acids, are extracted from plants, are liquid at room temperature, and are considered the "good dietary fats." Unsaturated fatty acids have at least one double bond between carbon atoms in the hydrocarbon chain. Thus, they have fewer hydrogen atoms.

LIPID FUNCTIONS

Lipids serve many functions.

1. **Energy storage:** 1 gram of any lipid will release 9 calories per gram when burned in a calorimeter.

2. **Structural:** Phospholipids are a major component of the cell membrane.

3. **Endocrine:** Some lipids are hormones.

Proteins

- Proteins are polymers or polypeptides consisting of repeating units called amino acids joined by peptide bonds.
- Amino acids consist of a carboxyl group, an amine group, and a variable (R), all attached to a central carbon atom. The R group, or variable, differs with each amino acid.
- Proteins are complex macromolecules and are responsible for growth and repair.
- Dietary sources of proteins include fish, poultry, meat, and certain plants like beans and peanuts, which are high in protein.
- One gram of protein burned in a calorimeter releases 4 calories.
- Proteins consist of the elements S, P, C, O, H, and N.
- With only 20 different amino acids, cells can build thousands of different proteins.
- Enzymes are examples of proteins.
- Figure 4.9 is a sketch of two amino acids combining to form a **dipeptide**. A dipeptide is a molecule consisting of two amino acids connected by one peptide bond.

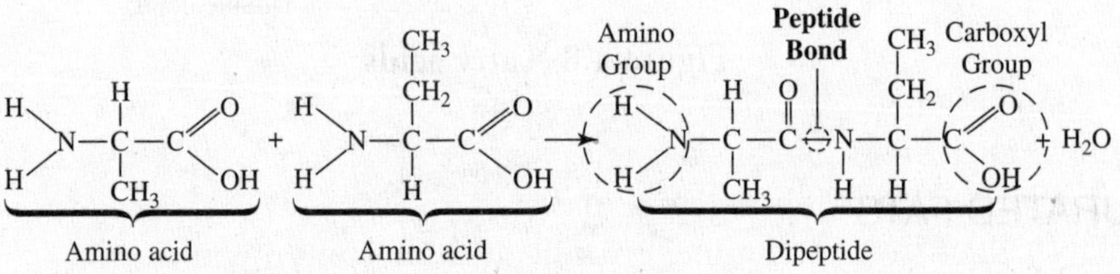

Figure 4.9

PROTEIN STRUCTURE

Proteins have many functions in living things. They act as enzymes, membrane channels, and hormones, to name a few. In every case, the function of the protein depends on its shape. The shape of a protein, in turn, is the result of four levels of structure: primary, secondary, tertiary, and quaternary.

Primary structure results from the sequence of amino acids that make up the protein chain.

Secondary structure results from the hydrogen bonding within the molecule. The helical nature of many proteins is the result of hydrogen bonding.

Tertiary structure is the intricate, three-dimensional shape or conformation of a protein and most directly determines the way it functions and its specificity. Enzymes **denature** (lose their natural shape) in high temperatures or adverse pH. When a protein/enzyme denatures, it cannot function because its tertiary structure has been altered beyond repair.

Quaternary structure refers to proteins that consist of more than one polypeptide chain. Hemoglobin, for example, exhibits quaternary structure because it consists of four chains; see Figure 4.10.

REMEMBER

Tertiary structure is directly responsible for the shape of a protein and how it functions.

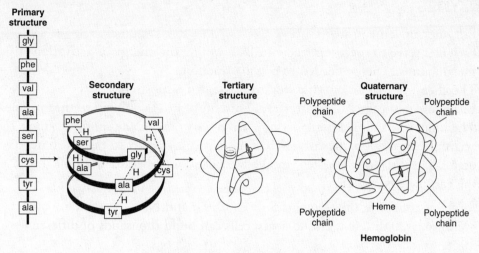

Figure 4.10 Protein structure

Nucleic Acids

The nucleic acids are deoxyribonucleic acid (DNA) and ribonucleic acid (RNA). They carry hereditary information.

- They are polymers (chains of repeating units) of **nucleotides.**
- A single nucleotide consists of a phosphate, a 5-carbon sugar (either deoxyribose or ribose), and a nitrogenous base.
- In DNA the bases are adenine, cytosine, guanine, and thymine.
- In RNA the bases are adenine, cytosine, guanine, and uracil.
- Adenine and guanine are **purines.**
- Cytosine, thymine, and uracil are **pyrimidines.**

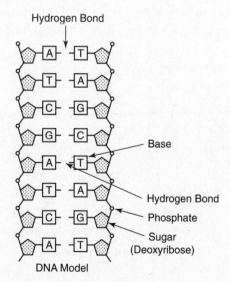

Figure 4.11 DNA molecule

ENZYMES

- Enzymes are large proteins.
- Enzymes serve to speed up reactions by lowering the **energy of activation** (E_a), the amount of energy needed to begin a reaction.
- The chemical that an enzyme works on is called a substrate.
- Enzymes are specific. In Figure 4.12, only substrate *A* will bind to the enzyme.
- The induced-fit model describes how enzymes work. As the substrate enters the active site, it induces the enzyme to alter its shape slightly so the substrate fits better. The old "lock and key" model was abandoned because it implied that the enzyme never changes.

> **INTERESTING FACT**
>
> Some people cannot digest dairy products because they lack the enzyme lactase.

Figure 4.12

- Enzymes are not degraded during a reaction and are reused.
- Enzymes are named after their substrate, and the name ends in the suffix "ase." For example, sucrase is the name of the enzyme that hydrolyzes sucrose, and lactase is the name of the enzyme that hydrolyzes lactose.
- Enzymes function with the assistance from **cofactors** (minerals) or **coenzymes** (vitamins).
- The efficiency of the enzyme is affected by temperature and pH. Average human body temperature is 37°C, near optimal for human enzymes. If body temperature rises above 40°C, the enzymes will stop functioning, as shown in Figure 4.13.

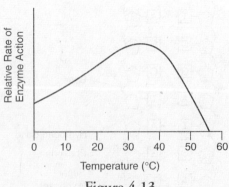

Figure 4.13

- As enzymes denature, they lose their unique shape and their ability to function.

• Gastric enzymes become active at low pH, when mixed with stomach acid. In constrast, intestinal amylase works best in an alkaline environment; see Figure 4.14.

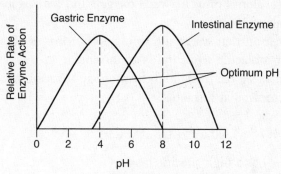

Figure 4.14

PRIONS—PROTEINS THAT CAUSE DISEASE

Prions are infectious proteins that cause several brain diseases, including mad cow disease. A prion is a misfolded version of a protein normally found in the brains of mammals. If a prion gets into a normal brain, it causes all the normal proteins to misfold in the same way.

MULTIPLE-CHOICE QUESTIONS

1. A solution with a pH of 2 is _____ times more acidic than one with a pH of 5.

   (A) 3
   (B) 10
   (C) 100
   (D) 1,000
   (E) 10,000

2. All of the following are correct about enzymes EXCEPT

   (A) the mechanism by which enzymes work is known as lock and key
   (B) they are proteins
   (C) they denature at high temperatures
   (D) they are assisted by vitamins and minerals
   (E) enzymes are not degraded during a reaction

3. All of the following are correct about water EXCEPT

   (A) water is a molecule
   (B) there is little attraction between water molecules
   (C) the covalent bonds between oxygen and hydrogen are polar or unbalanced
   (D) the reason that water and lipids do not mix is because water is a polar molecule while lipids are nonpolar
   (E) water has a relatively high heat of vaporization because of strong intermolecular attractions

4. Which is NOT a characteristic of water?

   (A) Water has a high specific heat.
   (B) Water has a high heat of vaporization.
   (C) Water exhibits strong cohesion tension.
   (D) Water is less dense than ice.
   (E) Water is known as a universal solvent

5. The pH of blood in humans

   (A) is lowest at birth and gradually increases with age up to a maximum level
   (B) is different for men and women
   (C) varies with the activity level of the individual
   (D) is highest at birth and gradually decreases to a minimum level
   (E) is normally 7.4 and resists change at all times

6. Which of the following is not a carbohydrate?

   (A) Glucose
   (B) Lactose
   (C) Insulin
   (D) Starch
   (E) Sucrose

7. Which of the following is not a polysaccharide?

   (A) Cellulose
   (B) Glycogen
   (C) Chitin
   (D) Glycerol
   (E) Starch

8. Which of the following is correctly matched?

    (A) Proteins—nucleotides
    (B) Lipids—glycerol
    (C) Carbohydrates—amino acids
    (D) DNA—glucose
    (E) None of the above is correctly matched.

Questions 9–15

Choose from these structural formulas below.

(A)

(B)

(C)

(D)

(E)

9. This is a monosaccharide.

10. **This is necessary for growth and repair of tissue.**

11. This combines with fatty acids to form lipids.

12. This is used as a quick energy source.

13. This is linked to cardiovascular disease.

14. This is an important part of any protein.

15. This consists of a sugar, a phosphate, and a nitrogenous base.

Questions 16–18

Match the description to the property of water.

     (A) Water exhibits strong cohesion tension.
     (B) Water has a high heat of vaporization.
     (C) Water has a high specific heat.
     (D) Ice is less dense than water.
     (E) Water is a universal solvent.

16. Water moves up tall trees because this is true.

17. Sweating is a cooling process because of this characteristic of water.

18. Fish can live through the winter in a lake that has ice floating on the surface.

19. Isotopes differ from each other only in

     (A) the number of electrons
     (B) the number of protons
     (C) the number of neutrons
     (D) how they react chemically
     (E) the size of the atom

20. All of the following are correct about enzymes EXCEPT

     (A) enzymes are organic catalysts
     (B) enzymes lower the energy of activation
     (C) enzymes are assisted by cofactors
     (D) enzymes are affected by changes in temperature but not changes in pH
     (E) enzymes are larger than the substrates they work on

21. A polysaccharide found in plants whose function is storage is

     (A) starch
     (B) glycogen
     (C) chitin
     (D) glucagon
     (E) cellulose

22. Enzymes function because of their particular shape or conformation. Which level of protein structure is most directly responsible for the shape of a protein?

    (A) Primary
    (B) Secondary
    (C) Tertiary
    (D) Quaternary

23. The radioisotope I-131 is used to diagnose and treat diseases of the

    (A) brain
    (B) thyroid
    (C) pancreas
    (D) lungs
    (E) stomach

EXPLANATION OF ANSWERS

1. **(D)** The pH is a measure of the concentration of H^+ ions in solution. The difference in concentration increases 10-fold for each increase in pH. A solution with a pH of 1 has a H^+ concentration of 1×10^{-1} molar, while a solution of pH 2 has a H^+ concentration of 1×10^{-2} molar. pH 1 has 10 times more H^+ than a solution of pH 2 and 100 times more H^+ than pH 3.

2. **(A)** The lock and key concept was abandoned in favor of the induced-fit model. The induced-fit theory states that the substrate induces the enzyme to change its shape to accommodate the substrate. All the other statements about enzymes are correct.

3. **(B)** There is strong hydrogen attraction between the oxygen and hydrogen molecules of adjacent water molecules.

4. **(D)** Ice is less dense than water. This is the reason that ice floats. All the other choices are true statements about water.

5. **(E)** The pH of blood for all humans is maintained at 7.4. This is an example of how the body maintains homeostasis or internal stability.

6. **(C)** Insulin is a protein. All the rest are carbohydrates. Glucose is a monosaccharide. Lactose and sucrose are disaccharides, and starch is a polysaccharide.

7. **(D)** Glycerol is a component of a lipid. One molecule combines with three fatty acids to form a lipid.

8. **(B)** A lipid consists of one glycerol molecule plus three fatty acid molecules.

9. **(A)** Glucose is a monosaccharide. Its chemical formula is $C_6H_{12}O_6$.

10. **(B)** Proteins are involved with growth and repair. They are polymers consisting of chains of amino acids.

11. **(D)** A lipid is a complex molecule that consists of one glycerol molecule plus three fatty acid molecules.

12. **(A)** Glucose is an energy source. Like all monosaccharides, 1 gram of glucose releases 4 calories.

13. **(C)** Saturated fatty acids are animal fats and are linked to cardiovascular disease.

14. **(B)** Proteins are polymers consisting of chains of amino acids.

15. **(E)** Nucleotides are the building blocks of the nucleotides, DNA and RNA.

16. **(A)** Water is pulled up a tree by a combination of cohesion tension and transpirational pull. As one molecule of water leaves the leaf by transpiration, another is drawn in at the roots.

17. **(B)** Because a relatively large amount of energy is required for evaporation due to water's high heat of vaporization, large amounts of energy in the form of heat are released from the body when sweating.

18. **(D)** Ice is less dense than water and floats on the surface of a lake in winter. Assuming the lake is not solidly frozen, life can exist underneath the ice year-round because it is insulated from severe conditions by the layer of ice.

19. **(C)** Isotopes are atoms of one element that vary only in the number of neutrons in the nucleus. The number of electrons is the same; therefore, chemical reactivity is the same. All samples of the same element have the same number of protons.

20. **(D)** Enzymes are organic catalysts that have a particular shape that determines how they function. Both temperature and pH can adversely affect the shape and functioning of an enzyme.

21. **(A)** Starch stores energy in plant cells; glycogen stores energy in animal cells. In humans, glycogen is found in the liver and muscles. Cellulose and chitin serve to support plants and animals, respectively. Glucagon is a hormone from the liver that elevates blood sugar.

22. **(C)** Tertiary structure is the three-dimensional conformation of the protein. It results from intramolecular forces of attraction other than hydrogen bonding. A protein denatures because the tertiary structure has changed.

23. **(B)** Radioactive iodine can be used both to diagnose and to treat thyroid diseases.

The Cell

CHAPTER 5

- Cell theory
- Comparison of prokaryotes and eukaryotes
- Structure of plant and animal cells
- Transport into and out of cells
- Life functions
- Tools to study cells

Anton van Leeuwenhoek invented the first microscope, a very fine magnifying lens, in the seventeenth century in Holland. He was the first person to observe living things under a microscope. In about 1665, Robert Hooke developed a microscope that enabled him to study cork tissue. He coined the term "cell." The German botanist Matthias Schleiden (1838), studying living tissue using these newly developed microscopes, concluded that all plants are made of cells. About the same time, Theodor Schwann (1839) stated that all animals are made of cells. Rudolf Virchow (1855), a pathologist studying cell reproduction, summarized his years of research by stating, "Where a cell exists, there must be a pre-existing cell." The work and discoveries of all these scientists is responsible for one of the fundamental theories of biology, the cell theory.

CELL THEORY

Modern cell theory states:

- All living things are composed of cells
- Cells are the basic unit of all organisms
- All cells arise from preexisting cells

Most plant and animal cells have diameters between 10–100 micrometers (μm). Many, though, like red blood cells, are very small, with a diameter of 8 μm. All cells are enclosed by a membrane that regulates the passage of materials between the cell and its surroundings. They also contain nucleic acid, which directs the cell's activities and controls inheritance.

Cells are divided into two varieties: prokaryotes and eukaryotes. **Prokaryotes** have no nucleus or other internal membranes. All bacteria are prokaryotes. **Eukaryotes** have a nucleus and are more complex cells. They make up every other form of life. Human cells are eukaryotic cells. Figure 5.1 shows a typical prokaryotic bacterial cell that lacks a nucleus and all membrane-bound internal structures. Figure 5.3 and Figure 5.4 show typical plant and animal cells.

Table 5.1 compares prokaryotic and eukaryotic cells.

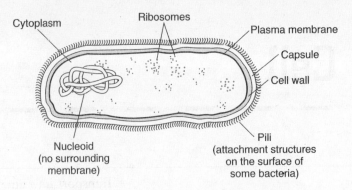

Figure 5.1 Typical prokaryotic cell

TABLE 5.1

Comparison of Prokaryotes and Eukaryotes

Prokaryotes	Eukaryotes
No membrane-bound organelles such as a nucleus	Contain distinct organelles surrounded by membranes, such as nucleus and mitochondria
Contains a single, circular chromosome	Chromosomes are wrapped with special proteins called histones; human body cells contain 46 chromosomes in each nucleus
Ribosomes are small	Ribosomes are larger
Respiration can be either aerobic or anaerobic	Respiration is mostly aerobic
Cytoskeletal elements, such as microfilaments, are absent	Cytoskeletal elements, like microfilaments and microtubules, are present
Most are unicellular	Some, like euglena and paramecium, are single celled; many are multicellular with specialized cell types, such as muscle, blood, and skin cells
Very small: 1–10 μm	Larger: 10–100 μm
Most have tough external cell walls	Most (except plant cells and protists) are surrounded by only a cell membrane

STUDY TIP

Form fits function.

Scientific studies using radioactive dating indicate that Earth is about 4.6 billion years old. All organisms on Earth are believed to have descended from a common ancestral prokaryotic cell about 3.5 billion years ago. According to the **theory of endosymbiosis**, eukaryotic cells containing organelles like mitochondria and chloroplasts, evolved when free-living prokaryotes took up permanent residence inside other larger prokaryotic cells, about 1.5 billion years ago. This was the origin of a complex eukaryotic cell with internal membranes that compartmentalized the cell, making it very efficient and leading to the evolution of all multicelled organisms.

Although all cells have many structures in common, they do not look alike. Many, in fact, are quite unique. A cell's form is dictated by its function. A nerve cell, for example, whose purpose is to conduct electrical impulses, is long and spindly.

Cells that make up a tough peach pit resemble square building blocks. The human body is made of approximately two hundred different eukaryotic cell types, each with a different function. Therefore, each cell type has a different form. Although different cell types have different functions and appearances, they all contain the same organelles; see Figure 5.2.

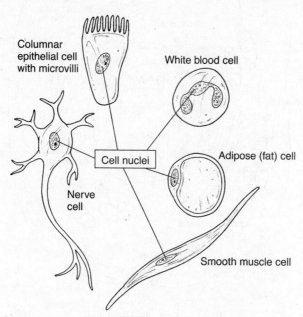

Figure 5.2 Five different eukaryotic cell types

STRUCTURES OF PLANT AND ANIMAL CELLS

Plant and animal cells have many cellular organelles in common, including ribosomes and mitochondria. However, they also have organelles unique to the cell type, such as cell walls in plant cells and centrioles in animal cells. Figure 5.3 and Figure 5.4 show typical plant and animal cells.

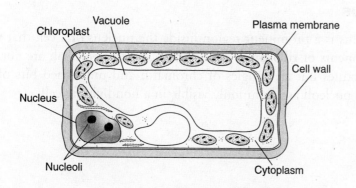

Figure 5.3 Plant cell

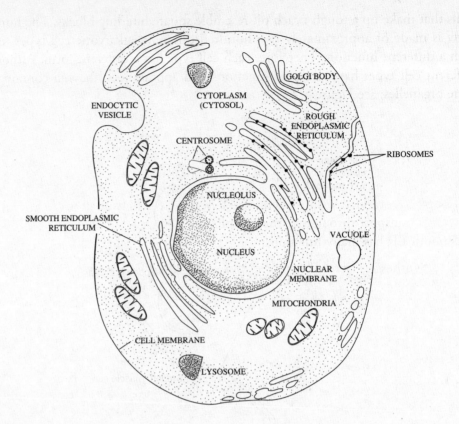

Figure 5.4 Animal cell

Nucleus

The nucleus contains chromosomes made of DNA that is wrapped with special proteins called **histones** into a **chromatin network**. Chromosomes contain genes, bits of DNA that code for polypeptides. The nucleus is surrounded by a selectively permeable double membrane or envelope that contains pores and allows for the transport of large molecules such as RNA out of the nucleus and into the cytoplasm.

Nucleolus

The **nucleolus** is a prominent region inside the nucleus of a cell that is not dividing. Components of **ribosomes** are synthesized here. Nucleoli are not membrane-bound structures but are tangles of chromatin and unfinished bits of ribosomes. One or two nucleoli are commonly visible in a nondividing cell.

Ribosome

This is the site of protein synthesis. Ribosomes are particles made of ribosomal RNA and protein. They are suspended freely in the cytoplasm or bound to endoplasmic reticulum. A single cell, such as a human liver cell, that produces large amounts of protein, contains millions of ribosomes.

Endoplasmic Reticulum

The **endoplasmic reticulum** (ER) is a system of membrane channels that traverse the cytoplasm. There are two varieties.

Rough ER studded with ribosomes. Therefore, it is the site of protein synthesis as well as transport throughout the cytoplasm.

Smooth ER has many functions:

1. Synthesizes steroid hormones and other lipids
2. Connects rough ER to the Golgi apparatus
3. Detoxifies the cell
4. Carbohydrate (glycogen) metabolism

Golgi Apparatus

The **Golgi apparatus** lies near the nucleus and consists of flattened sacs of membranes stacked next to each other (like a stack of pancakes) and surrounded by vesicles. They modify, store, and package substances produced in the rough endoplasmic reticulum. The Golgi apparatus also secretes these substances to other parts of the cell and to the cell surface for export to other cells.

Lysosome

A lysosome is a sac of hydrolytic (digestive) enzymes enclosed by a single membrane. It is the principal site of intracellular digestion. With the help of the lysosome, the cell continually renews itself by breaking down and recycling cell parts. Programmed cell death (**apoptosis**) is critical to the development of multicellular organisms and is carried out by a cell's own hydrolytic enzymes. Plant cells do not usually have lysosomes.

Mitochondrion

The **mitochondrion** (plural, mitochondria) is the site of cellular respiration. All cells have many mitochondria. A very active cell could have about 2,500 of them. Mitochondria consist of an outer double membrane and a series of inner membranes called **cristae**. Enzymes that are important to cellular respiration are embedded in the cristae membrane. Mitochondria contain their own DNA and can self-replicate. See Figure 5.5. (Remember, they were free-living prokaryotes several billion years ago.)

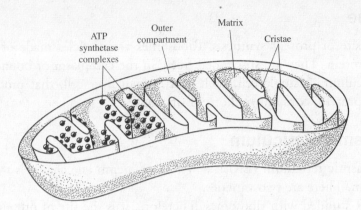

Figure 5.5 Mitochondrion

Vacuole

Vacuoles are single, membrane-bound structures that store substances for the cell. Freshwater protista, like paramecium and amoeba, have **contractile vacuoles** that pump excess water out of the cell. Plant cells and human fat (adipose) cells have large central vacuoles for storage.

Vesicle

Vesicles are tiny vacuoles. They are found in many places in cells, including the axon of a neuron, where they release neurotransmitter into a synapse.

Plastids

Plastids have a double membrane and are found only in plants and algae. There are three types.

1. **Chloroplasts** (Figure 7.1) are green because they contain chlorophyll. They are the sites of photosynthesis. In addition to a double outer membrane, they have an inner one that forms a series of structures called **grana**. The grana lie in the **stroma**. Chloroplasts, like mitochondria, contain their own nuclear material and can self-replicate. (Remember, they were free-living prokaryotes several billion years ago.)

2. **Leucoplasts** are colorless and store starch. They are found in roots, like turnips, or in tubers, like potatoes.

3. **Chromoplasts** store carotenoid pigments and are responsible for the red-orange-yellow coloring of carrots, tomatoes, daffodils, and many other plants. These bright pigments in petals attract pollinating insects to flowers.

Cytoskeleton

The cytoskeleton of a cell is a complex network of protein filaments that extends throughout the cytoplasm and gives the cell its shape and enables it to move. The cytoskeleton includes two types of structures.

1. **Microtubules** are thick hollow tubes that make up the cilia, flagella, and spindle fibers.

2. **Microfilaments** are made of the protein actin and help support the shape of the cell. They enable

   - Animal cells to form a cleavage furrow during cell division
   - Amoeba to move by sending out pseudopods
   - Skeletal muscles to contract by sliding along myosin filaments

Centrioles and Centrosomes

Centrioles and centrosomes lie outside the nuclear membrane and organize the spindle fibers required for cell division. Only animal cells have centrioles and centrosomes. Two **centrioles**, at right angles to each other, make up one **centrosome**. Centrioles and spindle fibers have the same structure. As shown in Figure 5.6, they consist of 9 triplets of **microtubules** arranged in a circle. (See right side of Figure 5.6.)

> **REMEMBER**
>
> Centrioles = 9 triplets

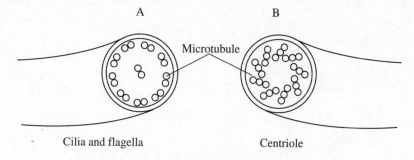

A   B

Microtubule

Cilia and flagella   Centriole

Figure 5.6

> **REMEMBER**
>
> Cilia and flagella = 9 + 2

Cilia and Flagella

Cilia and flagella have the same internal structure; both are made of microtubules. The only structural difference is in the length; cilia are short, and flagella are long. Figure 5.6 shows that both consist of 9 pairs of microtubules organized around 2 singlet microtubules. (See left side of Figure 5.6.)

Cell Wall

The cell wall is one structure not found in animal cells. Cell walls of fungi consist of **chitin**, while plants and algae have cell walls made of **cellulose**. In plant cells, the primary cell wall is immediately outside the plasma membrane. Some plant cells produce a secondary cell wall outside the primary cell wall. When a plant cell divides, a thin gluey layer is formed between the two cell walls, which becomes the **middle lamella** and which keeps the two daughter cells attached.

Cytoplasm and Cytosol

The entire region between the nucleus and plasma membrane is called **cytoplasm**. **Cytosol** refers to the semiliquid portion of the cytoplasm. In eukaryotic cells, organelles are suspended in the cytosol and get carried around the cell as the cytoplasm cycles around the cell, a process called **cyclosis**.

Cell or Plasma Membrane

The cell or plasma membrane is a selectively permeable membrane that controls what enters and leaves the cell. It is described as a fluid mosaic because it is made of many small particles that are able to move around in order to control substances entering and leaving the cell. The plasma membrane consists of a phospholipid bilayer with proteins dispersed throughout. Molecules of cholesterol are embedded within the membrane, making it less fluid and more stable. The external surface of the plasma membrane has carbohydrates attached to it that are important for cell-to-cell recognition, as can be seen in Figure 5.7.

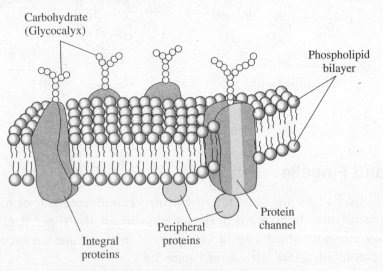

Figure 5.7 Plasma membrane

An average cell membrane consists of 60 percent protein. These proteins provide a wide range of functions for the cell. Some membrane proteins, like **ATP synthetase**, act as an enzyme. Some, like those involved in the **sodium-potassium pump**, transport ions into and out of cells.

Table 5.2 shows the differences between plant and animal cells.

TABLE 5.2

Differences Between Animal and Plant Cells

Animal Cells	Plant Cells
Centrioles and centrosomes	No centrioles or centrosomes
No chloroplasts and other plastids	Chloroplasts and other plastids
Small vacuoles	Large central vacuoles
Plasma membrane only	Cell walls in addition to plasma membrane
Lysosomes	No lysosomes

TRANSPORT INTO AND OUT OF THE CELL

Here is some important vocabulary for a discussion about transport.

1. **Selectively permeable.** A characteristic of a living membrane. The substances that pass through a selectively permeable membrane change with the needs of a cell. For example, the axon membrane of a nerve cell consists of gated channels that open and close to allow specific ions to pass through only when triggered by a certain stimulus.

2. **Solvent.** The substance that does the dissolving.

3. **Solute.** The substance that dissolves.

4. **Hypertonic.** Having a greater concentration of solute than another solution.

5. **Hypotonic.** Having a lower concentration of solute than another solution.

6. **Isotonic.** Two solutions containing equal concentrations of solute.

Passive Transport

Passive transport is the movement of molecules down a concentration gradient from a region of higher concentration to a region of lower concentration. Passive transport NEVER requires energy. It occurs either by **diffusion** (simple diffusion or facilitated diffusion) or by osmosis.

SIMPLE DIFFUSION

Simple diffusion is merely the movement of particles from a higher concentration to a lower concentration. The steeper the gradient, the faster the rate of diffusion. Earthworms "breathe" as oxygen from the air is absorbed by simple diffusion across their moist skin into capillaries directly beneath the skin. Humans obtain oxygen by simple diffusion across moist membranes in air sacs, called alveoli, in our lungs.

DON'T BE FOOLED

Facilitated diffusion does not require ATP/energy.

FACILITATED DIFFUSION

Facilitated diffusion relies on special protein membrane channels to assist in transporting specific substances across a membrane. For example, the normal functioning of a neuron requires calcium ions to be transported by facilitated diffusion through calcium ion channels within the axon membrane. Figure 5.8 shows a protein channel.

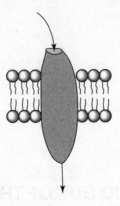

Figure 5.8

OSMOSIS

Osmosis is the diffusion of water across a membrane. Water flows down a gradient toward a region with high solute concentration. For example, in Figure 5.9, cell A has more solute than cell B. Therefore, cell A is hypertonic to cell B and cell B is hypotonic to cell A. Water will flow toward the region of higher concentration of solute, from cell B to cell A.

REMEMBER

Water diffuses toward the hypertonic area.

Diffusion of water

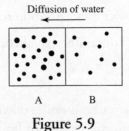

A   B

Figure 5.9

Figure 5.10 is a diagram of a cell in a hypertonic solution. Water will leave the cell, causing the cell to shrink. This cell shrinking is known as **plasmolysis**. In class, you may have carried out an experiment where you dropped a solution of 5 percent sodium chloride onto a living cell (such as elodea). Doing this caused the cell to shrink.

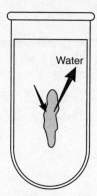

Figure 5.10

Cell in a hypertonic solution

Figure 5.11 is a diagram of a cell in a hypotonic solution. Water flows into the cell. This causes an animal cell to burst. If the cell is a plant cell, the cell wall will prevent the cell from bursting. Plant cells merely swell or become **turgid**. This turgid pressure is what keeps vegetables like celery or green peppers crisp. If a plant loses too much water (dehydrates), it loses **turgor pressure** and wilts.

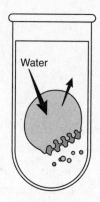

Figure 5.11

Cell in a hypotonic solution

Figure 5.12 shows a cell in an isotonic solution. Water diffuses in and out, but there is no net change in the cell. Solutions used to wash contact lenses or saline solutions used to wash out your eyes must be isotonic to protect your delicate body cells so as not to cause any damage.

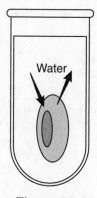

Figure 5.12

Cell in an isotonic solution

Active Transport

Active transport is the movement of molecules against a gradient, which requires energy, usually in the form of ATP. There are many examples of active transport in the biology course you studied.

The contractile vacuole in freshwater protista like paramecia and amoeba pumps out excess water that diffuses inward because the organisms live in an environment that is hypotonic.

Exocytosis is the active release of molecules from a cell. A good example is found in the synapse of nerve cells. Vesicles containing a neurotransmitter such as acetylcholine release their contents into the synapse in order to pass an impulse to another cell.

Pinocytosis, also called cell drinking, is the uptake of large, dissolved molecules. The plasma membrane invaginates around tiny particles and encloses them in a **vesicle**, as shown in Figure 5.13.

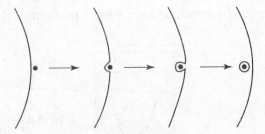

Figure 5.13 Pinocytosis

Phagocytosis is the engulfing of large particles or even small organisms by **pseudopods**. As shown in Figure 5.14, the cell membrane wraps around the particles and encloses them, forming a vacuole. This is the way human white blood cells engulf bacteria and also the way in which amoeba gain nutrition.

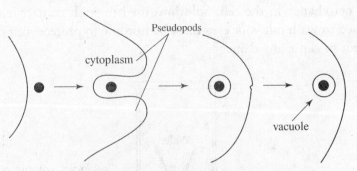

Figure 5.14 Phagocytosis

Receptor-mediated endocytosis enables a cell to take up large quantities of very specific substances. Extracellular substances bind to specific receptors on the cell membrane and are drawn into the cell into vesicles, as seen in Figure 5.15. This is the way in which body cells take up cholesterol from the blood.

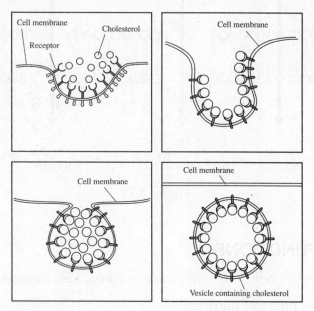

Figure 5.15 Receptor-mediated endocytosis

Membrane pumps carry particles or ions across a membrane against a gradient. The sodium-potassium pump in nerve cells, and seen in Figure 5.16, carries sodium (Na⁺) and potassium (K⁺) across the axon membrane to return the nerve to its resting state after an impulse has passed.

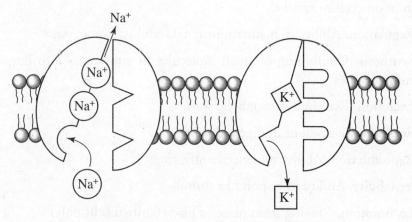

Figure 5.16 Sodium-potassium pump

Figure 5.17 shows an overview of passive and active transport.

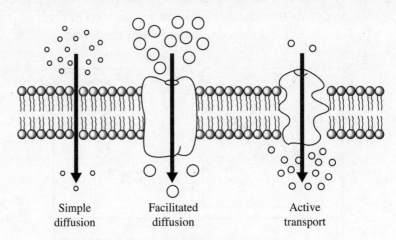

| Simple diffusion | Facilitated diffusion | Active transport |

Figure 5.17

THE LIFE FUNCTIONS

A characteristic of all cells is that they carry out certain life processes. They include:

1. **Ingestion.** Intake of nutrients

2. **Digestion.** Enzymatic breakdown, hydrolysis, of food so it is small enough to be assimilated by the body

3. **Respiration.** Metabolic processes that produce energy (adenosine triphosphate or ATP) for all the life processes

4. **Transport.** Distribution of molecules from one part of a cell to another or from one cell to another

5. **Regulation.** Ability to maintain internal stability, homeostasis

6. **Synthesis.** Combining of small molecules or substances into larger, more complex ones

7. **Excretion.** Removal of metabolic wastes

8. **Egestion.** Removal of undigested waste

9. **Reproduction.** Ability to generate offspring

10. **Irritability.** Ability to respond to stimuli

11. **Locomotion.** Moving from place to place (animal cells only)

12. **Metabolism.** Sum total of all the life functions

TOOLS AND TECHNIQUES TO STUDY CELLS

There are many tools and techniques to study cells. But, the main tool for studying cell structure (cytology) is the compound microscope. Besides the ability to magnify an image, another important characteristic of a good microscope is the measure of image clarity, known as **resolution**. The finest microscopes have both high magnification and excellent resolution. A toy microscope, which may enlarge an image 400×, has little resolving power, so the images are blurred.

Anton van Leeuwenhoek developed the first microscope in the seventeenth century. Today, compound light microscopes have been refined. Even the microscope you may have used in high school has fine resolution and good magnification. Figure 5.18 is a picture of a compound microscope. Learn the names of the working parts.

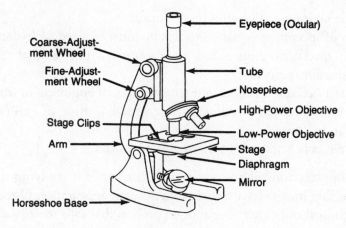

Figure 5.18

To determine the magnification, multiply the magnification of the **ocular lens** or eyepiece by the magnification of the **objective lens**. If the ocular has a magnification of 10×, which is customary, and the magnification of the objective lens is 40×, the total magnification is 400×.

When you use the microscope, remember that the image is upside-down and backward from the actual specimen you placed onto the slide. If you place a letter "e" onto the slide so that it appears the way you would read it in a book, the image will appear like it does in Figure 5.19 when you look in the microscope. Also, the higher the magnification you use, the darker the field will appear because you are viewing a much smaller area.

Figure 5.19

Today, the field of microscopy is very sophisticated. There are many different types of microscopes fashioned for different purposes.

1. Phase-contrast microscope
2. Transmission electron microscope
3. Scanning electron microscope

A phase-contrast microscope is a light microscope that enhances contrast. It is useful in examining living, unstained cells.

Electron microscopes use a beam of electrons, instead of a beam of light, to produce superior resolving power as well as magnification over 100,000×.

The transmission electron microscope (TEM) is useful for studying the interior of cells. The source of electrons is a tungsten filament within a vacuum column.

Although the TEM is very useful, there are some drawbacks:

- The tissue is no longer alive after processing.
- Preparation of specimens is elaborate. Tissue must be fixed, dehydrated, and sectioned on a special machine called a microtome, a process that requires many hours and much expertise.
- The TEM is a delicate machine and requires special engineers to maintain it.
- Specimens must be sliced so thin that only a small portion of a tissue sample can be studied at one time.
- The machine costs hundreds of thousands of dollars.

The scanning electron microscope (SEM) is useful for studying the surface of cells. The resulting images have a three-dimensional appearance. Once again, specimens are examined only after an elaborate process that kills the tissue.

Other Tools for Studying Cells

Another important tool used in the study of tissue is the **ultracentrifuge**. It enables scientists to isolate specific components of cells in large quantities by cell fractionation. By using this technique, cell components, such as mitochondria, can be studied under an electron microscope or analyzed biochemically.

First, tissue is mashed in a blender. The resulting liquid, called homogenate, is spun at high speed in an ultracentrifuge and separated into layers based on differences in density. The densest cell structures (nuclei) are forced to the bottom of the centrifuge tube. Less dense cell components (mitochondria and ribosomes) are layered above that. This is seen in Figure 5.20. If tissue is spun at high speed in a centrifuge tube, nuclei are forced to the bottom *first*, followed by mitochondria and ribosomes with clear liquid above the organelles.

Freeze fracture, also called freeze-etching, is a complex technique used to study details of membrane structure under an electron microscope. After preparation, only a cast of the original tissue is available to examine.

Tissue culture is a technique used to study the properties of specific cells in vitro (in the laboratory). Living cells are seeded onto a sterile culture medium to which a variety of nutrients and growth-stimulating factors have been added. Different cells require different growth media. Cell lines can be grown in culture for years provided great care is taken with them. While the cells are growing, they can be examined unstained under a phase-contrast light microscope.

DON'T FORGET

Specimens observed under the EM are not alive.

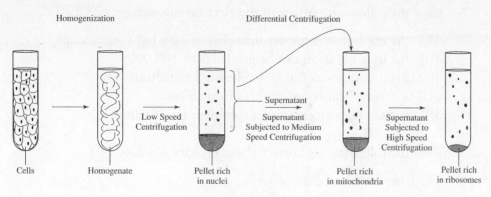

Figure 5.20

MULTIPLE-CHOICE QUESTIONS

<u>Questions 1–3</u>

Choose from the terms below.

    (A) Smooth endoplasmic reticulum
    (B) Nucleolus
    (C) Ribosomes
    (D) Lysosomes
    (E) Golgi apparatus

1. The site of protein production

2. Found in large quantities in white blood cells and other phagocytosing cells

3. The site of RNA production

4. The average eukaryotic cell has a diameter of about

    (A) 1.0 μm
    (B) 100 μm
    (C) 1000 μm
    (D) 1.0 nm
    (E) 10 nm

5. One difference between a plant cell and an animal cell is

    (A) only plant cells have chloroplasts while only animal cells have mitochondria
    (B) only plant cells have large vacuoles
    (C) only animal cells have nucleoli
    (D) only animal cells have plasma membranes
    (E) animal cells form a cleavage furrow when they divide

6. All of the following are true of the electron microscope EXCEPT

    (A) the resolution is greater than that using a light microscope
    (B) the magnification can be greater than $100,000\times$
    (C) living specimens can be studied in great detail
    (D) a narrow beam of electrons forms an image
    (E) it takes hours to prepare a specimen for viewing

7. All of the following are correct about prokaryotes EXCEPT

    (A) they have one chromosome
    (B) they contain ribosomes
    (C) they are smaller than eukaryotic cells
    (D) they can be either anaerobic or aerobic
    (E) they contain small mitochondria

8. The phase-contrast microscope

    (A) uses a beam of electrons to form an image
    (B) was developed by Robert Hooke
    (C) was developed by Anton van Leeuwenhoek
    (D) is commonly used to study genes
    (E) is useful in examining unstained, living tissue

9. Which would be best to study the structure of thousands of mitochondria?

    (A) Dissecting instruments and the scanning electron microscope
    (B) Dissecting instruments and the transmission electron microscope
    (C) Ultracentrifuge and the transmission electron microscope
    (D) Ultracentrifuge and the scanning electron microscope
    (E) Dissecting tools and the light microscope

10. What would occur if a plant cell were placed into a strong hypertonic solution?

    (A) Nothing would happen because the cell wall would prevent any movement of water or salt.
    (B) It would shrink because water would flow toward the more concentrated solution.
    (C) It would swell because salt would flow into the cell.
    (D) It would shrink because water would flow into the cell.
    (E) It would swell because water and salt would flow down the gradient and into the cell.

11. Which process requires energy?

    (A) Passive diffusion
    (B) Facilitated diffusion of calcium through a channel
    (C) Water flowing into a paramecium in a lake
    (D) A contractile vacuole removing water from an amoeba
    (E) Osmosis of water into a cell

12. The ability to respond to stimuli is

    (A) called irritability
    (B) called locomotion
    (C) characteristic of nerve cells only
    (D) characteristic of multicellular organisms only
    (E) regulation

13. All of the following are true of the smooth endoplasmic reticulum
    EXCEPT

    (A) it connects the rough endoplasmic reticulum to the Golgi apparatus
    (B) it detoxifies the cell
    (C) it synthesizes steroids
    (D) it manufactures proteins
    (E) it synthesizes lipids

14. Microtubules

    (A) form the cleavage furrow in a dividing cell
    (B) enable an amoeba to send out a pseudopod
    (C) are made of actin and help support the shape of a cell
    (D) are hollow tubes that make up cilia and flagella
    (E) are made of myosin and are key to skeletal muscle contraction

15. "Nine plus two" refers to the

    (A) configuration of microtubules in cilia and flagella
    (B) configuration of microtubules in centrioles and spindle fibers
    (C) configuration of microfilaments in cilia and flagella
    (D) configuration of microfilaments in centrioles and spindle fibers
    (E) configuration of myosin filaments in skeletal muscle

16. According to the theory of endosymbiosis,

    (A) autotrophic cells were the first to evolve
    (B) heterotrophic cells were the first to evolve
    (C) chloroplasts evolved from mitochondria
    (D) mitochondria evolved from chloroplasts
    (E) chloroplasts and mitochondria evolved when free-living prokaryotes
        permanently took up residence inside larger prokaryotic cells.

17. The circle below indicates the position of the letter X as seen in the field of your compound light microscope. To get the letter X in the center of the field, you would have to move the slide

    (A) to the right and down
    (B) to the left and up
    (C) to the left and down
    (D) to the right and up
    (E) it depends on the brand of microscope

18. All of the following are correct about mitochondria EXCEPT

    (A) they are the site of cellular respiration
    (B) they contain their own DNA
    (C) they are the site of photosynthesis
    (D) they can self-replicate
    (E) they contain an outer double membrane

19. All of the following are correct about plant cells EXCEPT

    (A) they usually contain one large vacuole
    (B) lysosomes contain hydrolytic enzymes for intracellular digestion
    (C) they store pigments in chromoplasts
    (D) they may have secondary cell walls in addition to primary cell walls
    (E) they have no centrioles

20. Which of the following involves growing living cells?

    (A) Freeze fracture
    (B) Freeze-etching
    (C) Cell fractionation
    (D) Tissue culture
    (E) Apoptosis

EXPLANATION OF ANSWERS

1. **(C)** Proteins are synthesized at the ribosomes.

2. **(D)** Lysosomes contain hydrolytic enzymes that digest cell parts or entire microbes.

3. **(B)** Nucleoli produce the RNA that makes up ribosomes.

4. **(B)** Most animal cells are about 100 μm in diameter. Prokaryotic cells are much smaller than eukaryotic cells.

5. **(E)** Plant cells form a cell plate when they divide but do not separate as animal cells do. Although only plant cells have chloroplasts, all eukaryotic cells contain mitochondria. The common plant cell contains one large vacuole. However, some animal cells, such as fat cells, contain only one large vacuole for storage. Both animal and plant cells contain nucleoli where RNA is synthesized. Both plant and animal cells have plasma membranes.

6. **(C)** Tissue studied under the electron microscope is not alive. Many hours are needed to prepare specimens for the electron microscope, killing the tissue in the process. The electron microscope, which uses a beam of electrons, has much better resolution and higher magnification than microscopes that use light to produce an image.

7. **(E)** Prokaryotes contain no mitochondria. They have no internal organelles made of membranes, such as mitochondria, nuclei, or vacuoles. They do, however, contain ribosomes, which are smaller than ribosomes in eukaryotic cells.

8. **(E)** The phase-contrast microscope is a light microscope that enhances contrast and can be used to visualize unstained, living cells. It is a modern scope. Van Leeuwenhoek lived in the 1600s, and Robert Hooke studied and named the cell in the 1800s. Genes cannot be seen with a phase-contrast microscope.

9. **(C)** To study the structure of thousands of mitochondria, you would first mash tissue in a blender and then spin it in an ultracentrifuge to isolate all the mitochondria. To study the structure of the mitochondria, which are very tiny, you would require a transmission electron microscope. A scanning electron microscope is used to study the surface of cells, not internal structures.

10. **(B)** Water flows toward a high concentration of salt, that is, toward a hypertonic solution. The cell would shrink, a process called plasmolysis. Hypertonic means contains a greater concentration of salt.

11. **(D)** A contractile vacuole pumps water against a gradient. Active transport requires energy and involves pumping substances against a gradient. Passive transport requires no energy. Examples of passive transport are simple diffusion, facilitated diffusion, and osmosis.

12. **(A)** Irritability is the ability to respond to stimuli. It is one of the life processes, like digestion or respiration. It is a characteristic of all living things, unicellular as well as multicellular.

13. **(D)** Proteins are synthesized by ribosomes, whether freely floating in the cytoplasm or attached to rough endoplasmic reticulum. All the other choices are true of smooth endoplasmic reticulum.

14. **(D)** Microtubules consist of protein that forms a hollow tube that makes up cilia, flagella, spindle fibers, and centrioles. The filaments responsible for muscle contraction are solid microfilaments consisting of actin and myosin.

15. **(A)** Cilia and flagella consist of nine pairs of microtubules arranged in a ring around two single microtubules located in the center. Spindle fibers and centrioles consist of nine triplets of microtubules arranged in a circle.

16. **(E)** According to the theory of endosymbiosis, small prokaryotic cells permanently took up residence inside larger prokaryotic cells about 3.5 billion years ago, forming mitochondria and chloroplasts. The heterotroph hypothesis states that the first organisms on Earth were heterotrophs. Choice A is not a true statement.

17. **(A)** The images in the microscope are all upside-down and backward. So if you wish the image to move up and to the left, you would have to move the slide down and to the right.

18. **(C)** Mitochondria are surrounded by a double membrane. They contain their own DNA and can self-replicate. They are the sites of the Krebs cycle, not the Calvin cycle.

19. **(B)** Plants do not contain lysosomes, animal cells do. The other statements are correct about plants.

20. **(D)** Freeze fracture and freeze-etching are the same thing. They produce a cast of the original tissue, which was vaporized during the procedure. Cell fractionation is accomplished by spinning a homogenate of tissue in a centrifuge. The cells separate based on differences in density. Apoptosis is programmed cell death. This destruction of cells is important in the development of many multicellular organisms. One example of apoptosis occurs during the metamorphosis of a tadpole to a frog. Lysosomes in the tail destroy these tail cells so that the tail will disappear.

Cell Division— Mitosis and Meiosis

CHAPTER 6

> • Definitions of mitosis and meiosis
> • Cell cycle
>
> • Stages of mitosis
> • Stages of meiosis

There are two types of cell division, mitosis and meiosis. **Mitosis** functions in the growth and repair of body cells. It produces two genetically identical daughter cells with the same chromosome number as the parent cell. Each daughter cell is diploid ($2n$), just like its parent cell.

Meiosis occurs in sexually reproducing organisms. It produces gametes (sperm and ova) with half the chromosome number of the parent cell. Each resulting cell is monoploid or haploid (n).

$$\text{Mitosis:} \quad 2n \nearrow \; 2n \atop \searrow \; 2n \qquad\qquad \text{Meiosis:} \quad 2n \nearrow \; n \atop \searrow \; n$$

STUDY TIP

Make sure you know the difference between mitosis and meiosis.

Any discussion about cell division must first inolve the structure of the chromosome. A chromosome consists of a highly coiled and condensed strand of DNA. A replicated chromosome consists of two **sister chromatids**, where one is an exact copy of the other. The **centromere** is a specialized region that holds the two sister chromatids together. Spindle fibers connect the centromere to the centrosome during cell division. Figure 6.1 shows a replicated chromosome.

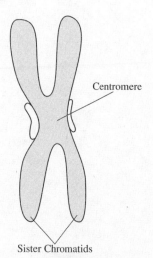

Centromere

Sister Chromatids

**Figure 6.1
A chromosome**

Figure 6.2 is of a cell containing four replicated chromosomes. Each chromosome contains two identical chromatids connected at a centromere.

Figure 6.2

THE CELL CYCLE

Living and dividing cells pass through a regular sequence of growth and division called the cell cycle. The cell cycle consists of five major stages: G_1, S, G_2 (which together are called interphase), mitosis, and cytokinesis. These phases are seen in Figure 6.3.

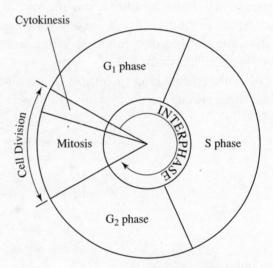

Figure 6.3 The cell cycle

Interphase

More than 90 percent of the life of a cell is spent in interphase. Most cells you have observed under a microscope are in this phase, including that shown in Figure 6.4.

During interphase, chromosomes replicate in preparation for cell division. One or more nucleoli become visible within the nucleus, and the nuclear membrane remains intact.

INTERPHASE
G₁
S
G₂

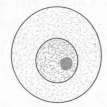

Figure 6.4 Interphase

Mitosis

Mitosis is the actual division of the nucleus. As seen in Figure 6.5, it is divided into four arbitrary phases: prophase, metaphase, anaphase, and telophase.

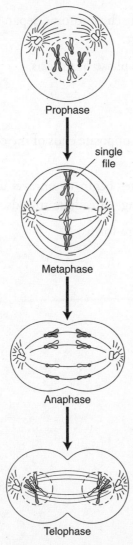

Prophase

single file

Metaphase

Anaphase

Telophase

Figure 6.5 Mitosis

PROPHASE

1. The nuclear membrane begins to disintegrate.

2. Strands of chromosomes begin to condense and become visible.

3. The nucleoli disappear.

4. **Spindle fibers** begin to form in the cytoplasm, extending from one centrosome to the other.

5. Prophase is the longest phase of mitosis.

METAPHASE

1. Chromosomes line up single file located on the equator or metaphase plate.

2. Centrosomes are at opposite poles of the cell.

3. Spindle fibers run from centrosomes to the centromeres of the chromosomes.

ANAPHASE

1. The centromeres of each chromosome separate, and spindle fibers pull the sister chromosomes apart.

2. Anaphase is the shortest phase of mitosis.

TELOPHASE

1. Chromosomes cluster at opposite ends of the cell, and the nuclear membrane re-forms.

2. Supercoiled chromosomes begin to unravel and to return to their pre-cell division condition as long, threadlike strands.

3. The nuclear membrane re-forms.

Cytokinesis

Cytokinesis is division of the cytoplasm. In animal cells, a cleavage furrow forms down the middle of the cell as the cytoplasm pinches inward and the two daughter cells separate from each other, as shown in Figure 6.6 (left). In plant cells, a **cell plate** forms down the middle of the cell. Daughter cells do not separate from each other. Instead, a sticky **middle lamella** cements adjacent cells together, as shown in Figure 6.6 (right).

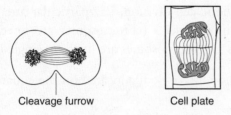

Cleavage furrow   Cell plate

Figure 6.6 Cytokinesis

MEIOSIS

Meiosis is a form of cell division in which cells having the diploid chromosome number ($2n$) produce gametes (sex cells) with the monoploid or haploid chromosome number (n). Meiosis occurs in two stages, meiosis I and meiosis II.

Meiosis I

1. This stage is also called reduction division.

2. **Synapsis** and **crossing-over** occur. During synapsis, chromosomes pair up precisely with their homologue so that crossing-over can occur. Crossing-over is the process in which homologous chromatids exchange genetic material. Crossing-over is important because it ensures greater variety in the gametes.

3. Homologous chromosomes then separate.

4. Next, chromosomes line up randomly on the equatorial plate and assort or separate independently. This means that how one pair of chromosomes lines up and separates has no effect on how any other pair of chromosomes lines up and separates.

5. Each resulting gamete is genetically unique.

THIS IS TRICKY

Meiosis I
Homologous pairs
separate

Meiosis II
Sister chromatids
separate

Meiosis II

1. This stage is similar to mitosis but does not have a special name.

2. Sister chromatids separate.

3. This division maintains the haploid, or monoploid, number of chromosomes.

4. This phase completes the goal of meiosis—producing four genetically unique cells from one original mother cell.

Figure 6.7 shows meiotic cell division. Pay particular attention to the difference between metaphase I in meiosis I (chromosomes are lined up double file) and metaphase II in meiosis II (chromosomes are lined up single file).

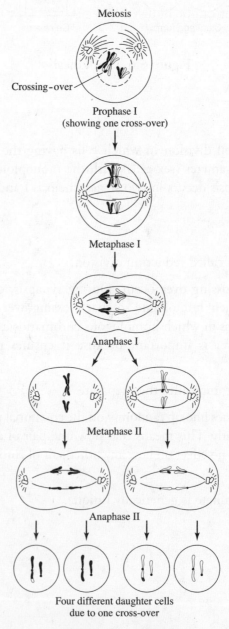

Figure 6.7 Meiosis

Information about spermatogenesis and oogenesis is presented in the chapter "Reproduction and Development."

The chapter "Plants" discusses meiosis and alternation of generations.

MULTIPLE-CHOICE QUESTIONS

Questions 1–6

    (A) Mitotic cell division
    (B) Meiotic cell division

1. The human liver can be induced to regenerate when liver cells become damaged and need to be replaced.

2. An embryo grown into a fetus and then a baby during a nine-month gestation period.

3. Human males begin to produce sperm at puberty.

4. A fertilized ovum undergoes rapid cell division immediately after it is fertilized by a sperm cell.

5. A cut on your skin heals.

6. Bacteria taken from a throat culture will grow in a petri dish.

7. Which of the following does NOT occur by mitosis?

    (A) Growth
    (B) Production of gametes
    (C) Repair
    (D) Development in the embryo
    (E) Cleavage furrow forms in animal cells

Questions 8–9

8. How many chromosomes are in this cell?

    (A) 0
    (B) 2
    (C) 3
    (D) 4
    (E) Cannot be determined

9. How many chromatids are in this cell?

    (A) 0
    (B) 2
    (C) 3
    (D) 4
    (E) Cannot be determined

10. In which stage of the life of a cell is the nucleolus visible?

    (A) Prophase
    (B) Anaphase
    (C) Telophase
    (D) Cytokinesis
    (E) Interphase

11. Which of the following is NOT found in plant cells?

    (A) Cell plate
    (B) Cleavage furrow
    (C) Middle lamella
    (D) Centromere
    (E) Chromatids

12. If a cell has 12 chromosomes at the beginning of meiosis, how many chromosomes will it have at the end of meiosis?

    (A) 6
    (B) 12
    (C) 24
    (D) 48
    (E) It varies with the species.

13. If a cell has 12 chromosomes at the beginning of mitosis, how many will it have at the end of mitosis?

    (A) 6
    (B) 12
    (C) 24
    (D) 48
    (E) It varies with the species.

14. All of the following are true of meiosis in plants EXCEPT

    (A) Crossing-over occurs during prophase.
    (B) There is no replication of chromosomes between meiosis I and meiosis II.
    (C) Spindle fibers are attached to the centriole.
    (D) Synapsis occurs during prophase.
    (E) The longest phase is prophase.

Questions 15–18

Match the event of mitosis with stages listed below.

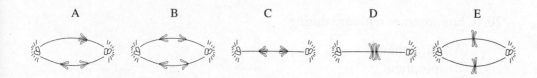

15. Which stage is metaphase I of meiosis?

    (A) A
    (B) B
    (C) C
    (D) D
    (E) E

16. Which stage is metaphase II of meiosis?

    (A) A
    (B) B
    (C) C
    (D) D
    (E) E

17. Which stage is anaphase II of meiosis?

    (A) A
    (B) B
    (C) C
    (D) D
    (E) E

18. Which stage is anaphase I of meiosis?

    (A) A
    (B) B
    (C) C
    (D) D
    (E) E

19. Chromosomes become visible for the first time during

    (A) interphase
    (B) prophase
    (C) metaphase
    (D) anaphase
    (E) telophase

20. Chromosomes replicate during

    (A) interphase
    (B) prophase
    (C) metaphase
    (D) anaphase
    (E) telophase

EXPLANATION OF ANSWERS

1. **(A)** Growth and repair are examples of mitosis.

2. **(A)** The cells of the embryo divide by mitosis.

3. **(B)** Sperm and ovum production is an example of meiotic cell division.

4. **(A)** The egg may have been formed by meiosis. Once it is fertilized, though, the growth and development of that zygote occurs by mitosis.

5. **(A)** Growth and repair are examples of mitosis.

6. **(A)** Growth and repair of any cell or tissue are examples of mitosis.

7. **(B)** Production of gametes (sperm and eggs) occurs by meiosis. This causes the chromosome number to be cut in half. The other choices are characteristic of mitotic cell division. A cleavage furrow forms in animal cells only.

8. **(B)** There are 2 replicated chromosomes in this cell but 4 chromatids. The number of chromosomes is determined by counting the number of centromeres, one per chromosome.

9. **(D)** There are 2 replicated chromosomes in this cell but 4 chromatids. The chromatids are the replicated chromosome strands and are joined at the centromere.

10. **(E)** Most of the life of the cell is spent in interphase when the chromosomes are threadlike and not visible under a light microscope. The nucleolus is not a real structure but actually threadlike chromosomes organized in a way that forms a spherelike structure. When the cell divides, chromosomes become condensed or supercoiled, and the nucleolus disappears.

11. **(B)** Animal cells develop a cleavage furrow as they undergo cytokinesis. Plant cells form a cell plate down the middle of the cell. Daughter plant cells do not separate from each other. Instead, sticky lamella cements adjacent plant cells together. All plant cells have chromosomes consisting of chromatids and centromeres.

12. **(A)** Meiosis cuts the chromosome number in half, from diploid *2n* to monoploid *n*. This division means that during fertilization, when two gametes fuse, the embryo has the correct chromosome number, *2n*. It does not vary with the species.

13. **(B)** The cells that result from mitotic cell division have the same number of chromosomes as the parent cell. It does not vary with the species.

14. **(C)** Plants do not have centrioles or centrosomes. Instead, the spindle fibers connect the chromosomes to a microtubule-organizing region. The longest phase of meiotic cell division is prophase because it is during that phase that synapsis and crossing-over occur. There is no replication of chromosomes between meiosis I and II because the chromosomes are already replicated double chromatids.

15. **(D)** During metaphase I of meiosis, homologous pairs line up on the metaphase plate in double file.

16. **(E)** During metaphase II of meiosis, homologous pairs line up on the metaphase plate in single file.

17. **(B)** During anaphase II, sister chromatids separate. The chromosome number remains the same.

18. **(C)** During anaphase I of meiosis, homologues separate. The chromosome number is cut in half.

19. **(B)** Most of the life of the cell is spent in interphase when the chromosomes are threadlike and not visible under a light microscope. When the cell divides, chromosomes become condensed or supercoiled and become visible. This supercoiling occurs during prophase.

20. **(A)** Chromosomal replication occurs during interphase. After replication, the chromosomes consist of two sister chromatids that are genetically identical to each other.

Photosynthesis

CHAPTER 7

- What is photosynthesis?
- Structure of the chloroplast
- Light and photosynthetic pigments
- Light-dependent reactions
- Light-independent reactions—the Calvin cycle
- Structure of the leaf

Photosynthesis is the process by which light energy is used to make glucose. Another way to describe it is to say that solar energy is converted into chemical energy because energy is stored in chemical bonds. Photosynthesis is carried out by all organisms in the plant kingdom as well as by algae in the Protista kingdom.

The general formula for photosynthesis is:

$$6CO_2 + 12H_2O \xrightarrow{\text{light}} C_6H_{12}O_6 + 6H_2O + 6O_2$$

Overall, this process is a **reduction** reaction because the carbon in carbon dioxide is gaining electrons from the hydrogen in water. (The definition of reduction is the gain of electrons.)

STRUCTURE OF THE CHLOROPLAST

The chloroplast is an organelle enclosed by a double membrane. It contains **grana** (consisting of layers of membranes called **thylakoids**)—where the light-dependent reactions occur, and **stroma**—where the Calvin cycle (light-independent reactions) occurs. Figure 7.1 is a sketch of a chloroplast.

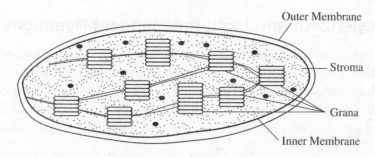

Figure 7.1 Chloroplast

Light and Photosynthetic Pigments

When light strikes an object, the light can be reflected, transmitted, or absorbed. Something that appears red reflects red light and absorbs light of all other colors. A green plant reflects green light and, therefore, cannot absorb green light as an energy source to make sugar. So if you were to shine only green light onto a green plant, it would die for lack of energy to make its own food.

Photosynthetic pigments absorb light energy and use it to carry out photosynthesis. Substances that absorb light in the **visible spectrum** (380 nm to 750 nm) are called pigments. Different pigments absorb light of different wavelengths. However, only the green photosynthetic pigment, **chlorophyll *a*,** can participate directly in the light reactions. Other pigments, called **antennae** or **accessory pigments**, assist in photosynthesis by capturing and passing on photons of light to chlorophyll *a* and expanding the range of light that can be used to produce sugar. These accessory pigments include **chlorophyll *b*, carotenoids,** and **phycobilins**. Chlorophyll *b* is green and absorbs all other wavelengths of light besides green. The carotenoids are yellow, orange, and red and are responsible for the color of carrots. The phycobilins are red and are found in red algae that live deep in the ocean where there is very little light.

Figure 7.2 is a graph showing the absorption spectrum for the photosynthetic pigments. Red and blue light are absorbed, while yellow and orange light are reflected. Only light that is absorbed can be used to power the making of sugar.

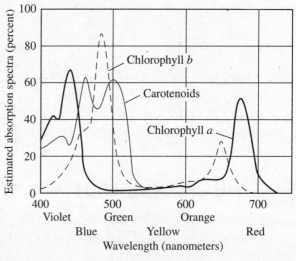

Figure 7.2

Light-Dependent and Light-Independent Reactions

There are two main processes of photosynthesis: the **light-dependent reactions,** or simply light reactions, and the **light-independent reactions,** or dark reactions. The light reactions require light directly, while the light-independent reactions do not. The dark reactions require only the products of the light reactions plus carbon dioxide, from the atmosphere, to synthesize sugar. Despite the names, both reactions occur only in the light.

LIGHT-DEPENDENT REACTIONS

The light-dependent reactions occur in the grana of chloroplasts within specialized membranes called thylakoids. Within these thylakoid membranes are hundreds of light-absorbing complexes called photosystems. Each photosystem consists of chlorophyll *a* and the accessory pigments chlorophyll *b* and carotenoids.

<table>
<tr><td>DON'T BE FOOLED</td></tr>
<tr><td>Both the light-dependent and light-independent reactions occur in the light.</td></tr>
</table>

There are two types of photosystems (I and II), which absorb light in slightly different wavelengths, thus expanding the wavelengths of light available for photosynthesis. In both photosystems, light is absorbed by chlorophyll *a*, causing excited electrons to escape into two electron transport chains where ATP is produced. This ATP will power the production of sugar in the light-independent reactions.

Here are some more facts about the light reactions:

- Electrons flow through the electron transport chains. Energy (**ATP**) is released by what is called *photo*phosphorylation since the energy to produce it comes from light energy.
- **Photolysis** occurs. Water is ripped apart to provide electrons to replace those lost by chlorophyll *a*.
- Oxygen molecules from the photolysis of water are released into the atmosphere as a waste product.
- In addition to replacing lost electrons to chlorophyll *a*, water also provides hydrogen for the dark reactions.
- NADP carries hydrogen from the photolysis of water to the dark reactions as NADPH.

Figure 7.3 is a sketch of the light-dependent reactions within the thylakoid membranes of chloroplasts. It shows the photosystems, electron transport chains, and the ATP synthetase channels where ATP is produced.

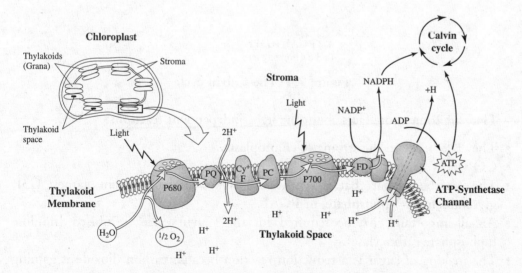

Figure 7.3 The light-dependent reactions

LIGHT-INDEPENDENT REACTIONS—THE CALVIN CYCLE

The function of the light-independent reactions is to produce sugar. Carbon dioxide, which the plant takes in through the **stomates**, combines with hydrogen carried from the light reactions by NADP to produce sugar (CH_2O). The incorporation of carbon dioxide into a sugar is called **carbon fixation**, and it occurs during the cyclical process known as the **Calvin cycle**, as shown in Figure 7.4. The energy for this process comes from the ATP that was produced in the light reactions.

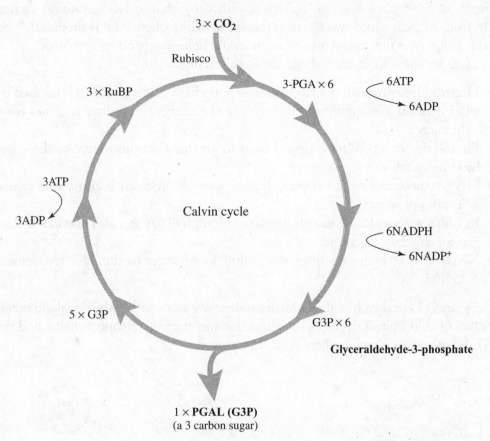

Figure 7.4 The Calvin cycle

Here are some more facts about the light-independent reactions:

- They take place in the stroma of chloroplasts.
- $CO_2 + H_2 \rightarrow$ sugar (CH_2O)
- The 3-carbon sugar that is synthesized is known by different names: **G3P**, **glyceraldehyde-3-phosphate**, or **PGAL**.
- An all-important enzyme required for the Calvin cycle is rubisco (ribulose biphosphate carboxylase).
- The making of sugar is a reduction reaction because carbon dioxide is gaining hydrogen.

Figure 7.5 shows where the light-dependent and Calvin cycle occur.

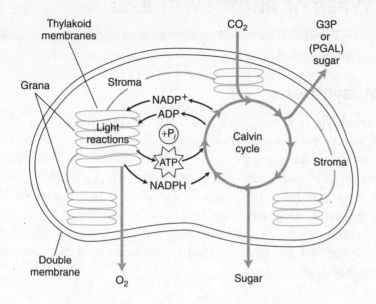

**Figure 7.5 Overview of light-dependent and
light-independent reactions**

STRUCTURE OF THE LEAF

Figure 7.6 is a cross section of a typical leaf. Observe that the palisade layer consists of tightly packed cells that contain chloroplasts. This is where the majority of photosynthesis occurs. Like the palisade layer, the cells of the spongy mesophyll contain chloroplasts and also carry out photosynthesis. However, the cells in the spongy mesophyll are less tightly packed and are surrounded by air spaces, which allow for the exchange of oxygen, carbon dioxide, and water vapor. The epidermis layer is clear and does not carry out photosynthesis. It protects the delicate underlying cells and allows light to pass into the leaf. Above the epidermis is a waterproof layer of **cutin** that minimizes excessive water loss. Guard cells control the opening and closing of the stomates, which allow for the exchange of gases with a minimum of water loss.

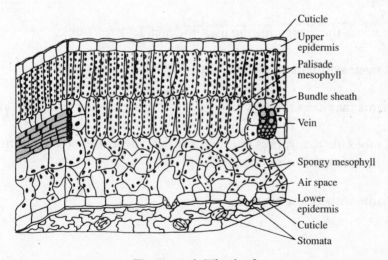

Figure 7.6 The leaf

OTHER TYPES OF PHOTOSYNTHESIS

Plants in dry environments carry out modified forms of photosynthesis. These alternatives, C-4 photosynthesis and CAM, allow the plants to retain more water.

C-4 Photosynthesis

C-4 photosynthesis is a modification for dry environments. C-4 plants exhibit modified anatomy and biochemical pathways, which enable them to minimize water loss and maximize sugar production. As a result of these modifications, C-4 plants thrive in hot and sunny environments where C-3 plants—more typical plants—would wilt and die. In C-4 plants, a series of steps precedes the Calvin cycle and pumps carbon dioxide deep into the leaf to bundle sheath cells. This alternate biochemical pathway is known as the Hatch-Slack pathway. The anatomical modification of the leaf of C-4 plants is called Kranz anatomy. Examples of C-4 plants are corn, sugarcane, and crabgrass.

CAM Plants

CAM stands for **crassulacean acid metabolism**. This is a form of photosynthesis that is another adaptation for dry conditions. These plants keep their stomates closed during the day and open at night, the reverse of how most plants behave.

MULTIPLE-CHOICE QUESTIONS

<u>Questions 1–5</u>

Choose from the terms below:

    (A) CAM
    (B) C-4
    (C) G3P
    (D) NADP
    (E) ATP

1. Provides the energy for the light-independent reactions

2. A sugar synthesized during photosynthesis

3. Plants that keep their stomates closed during the day and open at night

4. A molecule that carries H_2 from the light-dependent reactions to the light-independent reactions

5. Plants with Kranz anatomy

6. The oxygen that plants give off

    (A) comes from the light reactions of photosynthesis
    (B) comes from the light-independent reaction of photosynthesis
    (C) is a by-product of respiration
    (D) comes from carbon dioxide that plants absorb
    (E) is carried from the Calvin cycle to the light reactions by NADP

7. All of the following statements are correct about photosynthesis EXCEPT

    (A) water provides hydrogen for carbon fixation
    (B) the light-independent reactions occur in the stroma
    (C) the light-independent reactions do not require light and can occur
         at night
    (D) the light-dependent reactions occur in the thylakoid membranes
    (E) the thylakoid membranes are part of the grana

8. All of the following statements are correct about the light-dependent
   reactions of photosynthesis EXCEPT

    (A) carotenoids are accessory pigments
    (B) they take place in the grana
    (C) when light strikes chlorophyll *a*, electrons enter an electron
         transport chain
    (D) glyceraldehyde-3-phosphate, a 3-carbon sugar, is produced
    (E) they take place within the thylakoid membranes

9. All of the following statements about photolysis are correct EXCEPT

    (A) it occurs during the light-independent reaction
    (B) it releases oxygen
    (C) it releases hydrogen
    (D) it supplies electrons to chlorophyll *a*
    (E) it is the process of splitting apart molecules of water

10. Here is an absorption spectrum for an unknown substance. What color is the substance?

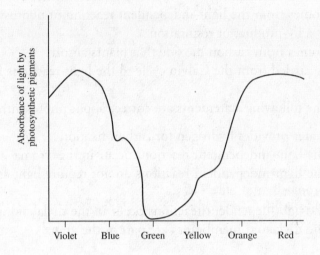

(A) Red
(B) Blue
(C) Green
(D) Yellow
(E) Violet

Questions 11–15

Indicate which of the following events occurs during

(A) Light-dependent reactions
(B) Light-independent reactions

11. Oxygen is released

12. Carbon gets incorporated into a sugar

13. ATP is produced

14. Electrons flow through an electron transport chain

15. Formation of NADPH

16. The sugar formed from photosynthesis is

    (A) ATP
    (B) glucose
    (C) pyruvic acid
    (D) G3P
    (E) sucrose

17. Which of the following is not a photosynthetic pigment?

    (A) Chlorophyll *a*
    (B) Chlorophyll *b*
    (C) Carotenoid
    (D) Bromothymol blue
    (E) Phycobilins

18. Carotenoids absorb all colors of light EXCEPT

    (A) green
    (B) blue
    (C) yellow
    (D) red
    (E) violet

19. The initial role of chlorophyll in photosynthesis is to

    (A) synthesize sugar
    (B) absorb light energy
    (C) fix carbon dioxide
    (D) synthesize carbon
    (E) convert ADP to ATP

20. Which of the following is an end product of both cellular respiration and the light-dependent reactions of photosynthesis?

    (A) glucose
    (B) ATP
    (C) CO_2
    (D) pyruvic acid
    (E) oxygen

EXPLANATION OF ANSWERS

1. **(E)** ATP is produced by photophosphorylation during the light reactions and is used to produce sugar during the Calvin cycle of the light-independent reactions.

2. **(C)** G3P stands for phosphoglyceraldehyde-3-phosphate. Sometimes it is called PGAL, phosphoglyceraldehyde. It is a sugar.

3. **(A)** CAM stands for crassulacean acid metabolism. It is a modification for a dry environment; the stomates close during the sunny and hot daytime.

4. **(D)** NADP is a molecule that ferries hydrogen from the light reactions occurring in the grana to the Calvin cycle, which occurs in the stroma.

5. **(B)** C-4 plants have evolved biochemical and anatomical modifications for dry environments. They exhibit Kranz anatomy and the Hatch-Slack pathways that enable a plant to keep its stomates closed more of the time than C-3 plants do. Thus, C-4 plants conserve more water.

6. **(A)** When light strikes chlorophyll *a*, electrons become excited and escape from chlorophyll *a*. Replacement electrons are supplied by electrons from water, not from carbon dioxide. The product of the Calvin cycle is sugar. Carbon dioxide is a by-product of respiration in both plants and animals. NADP carries hydrogen, not oxygen, from the light reactions to the Calvin cycle (light-independent reactions).

7. **(C)** The light-independent reactions do not require light. However, they require the products of the light-dependent reactions, ATP and hydrogen from NADPH. The light-dependent reactions occur in the grana, which consist of thylakoid membranes. The process of water breaking apart, called photolysis, provides electrons and protons for photosynthesis.

8. **(D)** Glyceraldehyde-3-phosphate (G3P), a 3-carbon sugar, is the product of the Calvin cycle, light-independent reactions, not the light-dependent reactions. The light-dependent reactions take place in the thylakoid membranes of the grana within chloroplasts. Chlorophyll *b* and carotenoids are antennae or accessory pigments.

9. **(A)** Photolysis is the spitting apart of water molecules into their components: oxygen and hydrogen. It occurs during the light-dependent reactions, not during the light-independent reactions. Electrons in chlorophyll *a* become excited when they absorb light energy and jump to electron transport chains in the thylakoid membrane. Water provides two things: electrons to replace those lost by chlorophyll *a* and hydrogen for production of sugar. In addition, oxygen is released into the environment.

10. **(C)** If light is absorbed, it is not reflected. Only reflected colors are seen. The graph shows that red and blue light are most absorbed and that green is most reflected. Therefore, the color of the substance is green.

11. **(A)** Oxygen is released from the photolysis of water during the light-dependent reactions.

12. **(B)** Carbon gets incorporated into sugar (carbon fixation) during the Calvin cycle, which occurs during the light-independent reactions.

13. **(A)** ATP is produced during the light-dependent reactions the same way it is produced during cell respiration, by chemiosmosis.

14. **(A)** Electrons flow through the electron transport chain during the light reactions. ATP and NADPH are produced.

15. **(A)** NADP gains hydrogen (is reduced to NADPH) during the light-dependent reactions and carries the hydrogen to the Calvin cycle for the light-independent reactions.

16. **(D)** G3P stands for glyceraldehyde-3-phosphate. It, not glucose, is the first stable carbohydrate produced by photosynthesis. ATP is a source of immediate energy, but it is not a sugar. Pyruvic acid is a 3-carbon compound that is a product of the anaerobic phase of respiration. Sucrose is a disaccharide that plants make from glucose.

17. **(D)** Bromothymol blue is a pH indicator commonly used in experiments, not a photosynthetic pigment. All the rest of the choices are pigments.

18. **(C)** We see whatever colors are reflected off an object. If you shine white light (which contains all colors of light) on carotenoids, you will see yellow-orange. This is because those colors of light are being reflected. All other colors of light are being absorbed.

19. **(B)** The first thing to happen during photosynthesis is that chlorophyll absorbs light. This light energy is then used to make ADP into ATP and ultimately to make sugar. Carbon is never synthesized during photosynthesis. It is fixed (combined with hydrogen to become part of a molecule) into sugar.

20. **(B)** ATP is produced during both the anaerobic and aerobic phases of cellular respiration as well as during the light-dependent reactions of photosynthesis. Glucose is a raw material of respiration, not a product. Carbon dioxide is a by-product of the Krebs cycle of aerobic respiration. Pyruvic acid is a product of the anaerobic phase of respiration. Oxygen is a by-product of photosynthesis, but required for aerobic respiration.

Cell Respiration

CHAPTER 8

• An overview of cell respiration	• Krebs cycle
• What is ATP?	• What are NAD and FAD?
• Structure of the mitochondria	• Electron transport chain
• Anaerobic respiration	• Oxidative phosphorylation
• Glycolysis	and chemiosmosis

Cell respiration is a series of oxidative reactions by which cells release energy from glucose and transfer it to molecules of **ATP** (adenosine triphosphate). Energy stored in ATP is immediately available for cellular activities such as contracting muscles, passing an impulse along a nerve, or pumping water by active transport.

Here is the equation for the complete aerobic respiration of one molecule of glucose:

$$C_6H_{12}O_6 + 6O_2 \quad \rightarrow \quad 6CO_2 + 6H_2O + energy$$

ATP—ADENOSINE TRIPHOSPHATE

ATP is the special high-energy molecule that stores energy for immediate use in the cell. It consists of adenosine (the nucleotide adenine + ribose) plus 3 phosphates. Figure 8.1 shows the structure of a molecule of ATP.

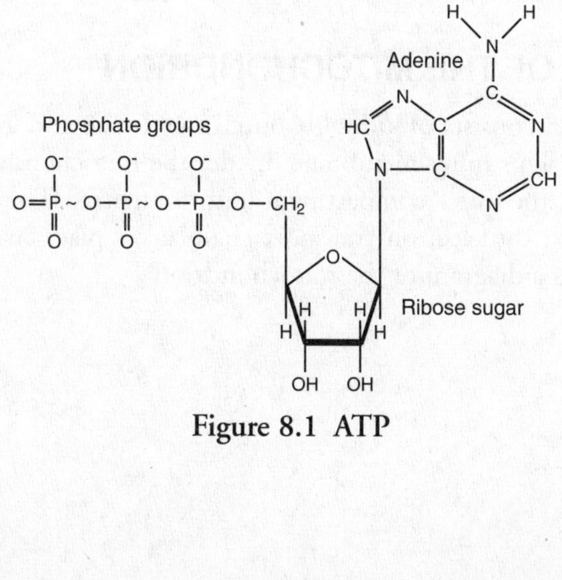

Figure 8.1 ATP

Cell respiration can be either aerobic or anaerobic. Table 8.1 shows the two different types and their component phases.

TABLE 8.1

Types of Cellular Respiration

	Anaerobic phase	Aerobic phase
Anaerobic Respiration	(1) Glycolysis (2) Alcoholic or lactic acid fermentation	None
Aerobic Respiration	Glycolysis	(1) Krebs cycle (2) Electron transport chain

The removal of one phosphate group from ATP results in the formation of a more stable and lower energy molecule, ADP. Energy is released as ATP converts to ADP. Energy is required to add a phosphate to ADP to produce ATP. Figure 8.2 shows this conversion.

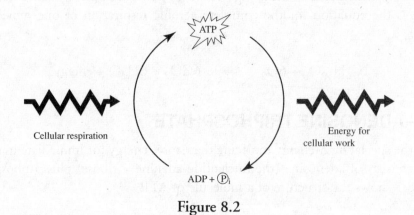

Figure 8.2

STRUCTURE OF THE MITOCHONDRION

The mitochondrion consists of a double outer membrane and a convoluted inner **cristae** membrane. This inner membrane divides the mitochondria into two internal compartments, the outer compartment and the matrix. The Krebs cycle takes place in the matrix; the electron transport chain takes place in the cristae membrane. Figure 8.3 is a diagram of the mitochondrion.

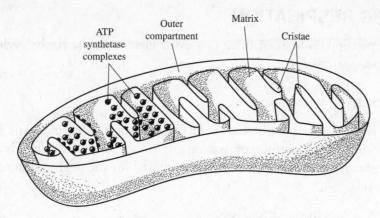

Figure 8.3 Mitochondrion

ANAEROBIC RESPIRATION

Anaerobic respiration or **fermentation** consists of the process known as **glycolysis** plus either **alcohol fermentation** or **lactic acid fermentation**. Anaerobic respiration originated billions of years ago when no free oxygen was availabale in Earth's atmosphere. Today it is the sole means by which anaerobic bacteria such as *Clostridium botulinum* (the bacterium that causes a form of food poisoning, botulism) release energy from food. Figure 8.4 shows the two types of fermentation, alcohol fermentation and lactic acid fermentation.

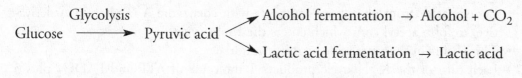

Figure 8.4

Alcohol Fermentation

Alcohol fermentation is the process by which certain cells convert pyruvic acid or pyruvate from glycolysis into ethyl alcohol and carbon dioxide in the absence of oxygen. The bread-baking industry depends on the ability of yeast to carry out fermentation and produce the carbon dioxide that causes bread to rise. The beer, wine, and liquor industry depends on yeast to ferment sugar into ethyl alcohol.

Lactic Acid Fermentation

Lactic acid fermentation occurs during strenuous exercise when the body cannot keep up with the increased demand for oxygen by skeletal muscles. Pyruvic acid produced by glycolysis converts to lactic acid and builds up in the muscles, causing fatigue and burning. The expression, "No pain, no gain" refers to the pain caused by lactic acid buildup in skeletal muscles. When blood restores proper oxygen levels, the muscle tissue reverts to the more efficient aerobic respiration. Lactic acid is removed from the muscles and carried to the liver, where it is converted back to pyruvic acid.

> **STUDY TIP**
>
> Fermentation does not produce any ATP.

AEROBIC RESPIRATION

Aerobic respiration consists of three processes: glycolysis, the **Krebs cycle**, and the **electron transport chain**.

Glycolysis

Glycolysis is the anaerobic phase of aerobic respiration. One molecule of glucose breaks apart into two molecules of pyruvate. Pyruvate, or pyruvic acid, is essentially one-half a glucose molecule and is the raw material for the next step in respiration, the Krebs cycle.

- Glycolysis occurs in the cytoplasm.
- Glycolysis is a complex, ten-step process, each step of which is controlled by a specific enzyme.
- Two molecules of ATP supply the energy of activation, the energy needed to begin the reaction.
- Glycolysis releases 4 ATP molecules, resulting in a net gain of 2 ATP
- **ATP** is produced by what is called **substrate level phosphorylation**, a process by which an enzyme transfers a phosphate group to ADP.
- 1 Glucose + 2 ATP → 2 Pyruvate + 4 ATP + 2 NADH (net gain 2 ATP)

The Krebs Cycle

The Krebs cycle, also known as the citric acid cycle, is the first stage of the aerobic phase of cellular respiration. It is shown in Figure 8.5.

- Pyruvic acid (from glycolysis) combines with coenzyme A (a vitamin A derivative) to form acetyl coA, which enters the Krebs cycle.
- This occurs in the matrix of the mitochondria.
- Each turn of the Krebs cycle produces 1 molecule of ATP and $FADH_2$ plus 3 molecules of NADH.
- The by-product is CO_2, which is exhaled.
- ATP is produced by **substrate level phosphorylation**, a process by which ATP is produced as a special enzyme moves a phosphate from a certain molecule to ADP.

NAD and FAD

NAD and FAD molecules are an important part of cell respiration.

- NAD and FAD are coenzymes that shuttle protons or electrons from glycolysis and the Krebs cycle to the electron transport chain.
- NAD^+ is the oxidized form; NADH is the reduced form. FAD^+ is the oxidized form; $FADH_2$ is the reduced form.

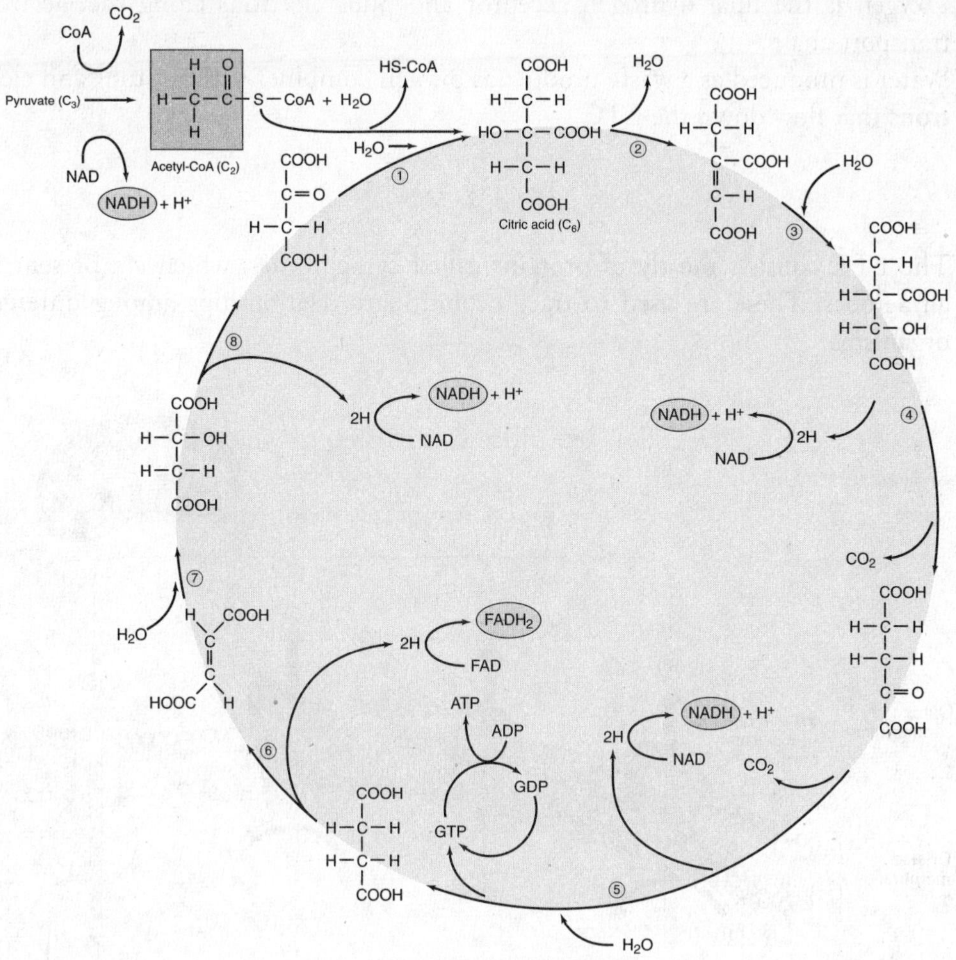

REMEMBER

NAD and FAD carry H^+ (protons) from the Krebs cycle to the electron transport chain.

Figure 8.5 Krebs cycle

The Electron Transport Chain

The electron transport chain (ETC) pumps protons across the cristae membrane of mitochondria in order to create a proton gradient. This gradient will power the phosphorylation of ADP to ATP during **oxidative phosphorylation**.

The electron transport chain makes no ATP directly but sets the stage for ATP production during oxidative phosphorylation. The ETC is complex. See Figure 8.6.

- The ETC uses the energy from the energy-releasing (exergonic) flow of electrons to pump protons across the cristae membrane in order to create a proton gradient.
- The ETC is a collection of molecules embedded in the cristae membrane of mitochondria.
- There are thousands of copies of the ETC in every mitochondrion due to the extensive folding of the cristae membrane.
- The ETC carries electrons delivered by NADH and FADH from glycolysis and the Krebs cycle to oxygen through a series of redox reactions. In a redox reaction, one atom gains electrons (**reduction**) while one atom loses electrons (**oxidation**).

- Oxygen is the final hydrogen acceptor and pulls electrons along the electron transport chain.
- Water is produced as a waste product as oxygen combines with protons and electrons that flow down the ETC.

$$\frac{1}{2}\ O_2 + H_2\ \rightarrow\ H_2O$$

- The ETC consists mostly of proteins called cytochromes, which are present in all aerobes. These are used to trace evolutionary relationships among different organisms.

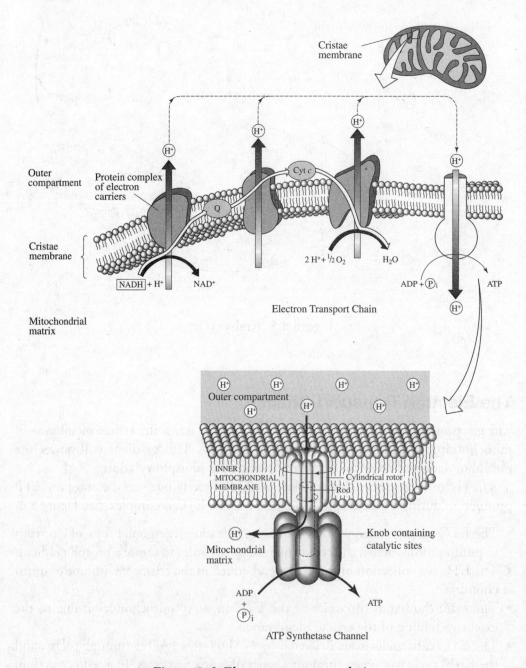

Figure 8.6 Electron transport chain

Oxidative Phosphorylation and Chemiosmosis

Most of the energy (ATP) produced during cell respiration occurs in the mitochondria by a process known by the general name, oxidative phosphorylation. This process was proposed in 1961 by Peter Mitchell, who named it **chemiosmosis.** According to the Mitchell hypothesis, chemiosmosis uses potential energy stored in a proton (H^+) gradient to phosphorylate ADP and produce ATP (ADP + P $\rightarrow$ ATP), just as water flowing over a dam can generate electricity.

Here are the important points about oxidative phosphorylation:

• The steep proton (electrochemical) gradient that was created by the ETC between the outer compartment and inner matrix in mitochondria is what provides the energy to produce ATP.

• Protons cannot diffuse through the cristae membrane but cross it through ATP synthetase channels.

• In chemiosmosis, protons can flow through only the special ATP synthetase channels and generate energy to phosphorylate ADP into ATP. It is similar to how a hydroelectric plant converts the enormous potential energy of water flowing through a dam to turn turbines and generate electricity.

• Hypothetically, each proton carried by an NAD molecule to the ETC produces 3 ATP molecules, while the protons carried by a FAD molecule produces 2 ATP molecules.

During respiration, most energy flows in this sequence:

Glucose $\rightarrow$ NAD and FAD $\rightarrow$ Electron transport chain (chemiosmosis) $\rightarrow$ ATP

Both Table 8.2 and Figure 8.7 provide overview of aerobic respiration.

TABLE 8.2

Phases of Respiration

Process	Location	Net Amount of ATP Produced
Glycolysis	Cytoplasm	2 ATP
Krebs cycle	Mitochondria	2 ATP
Electron transport chain (oxidative phosphorylation)	Mitochondria	32 ATP

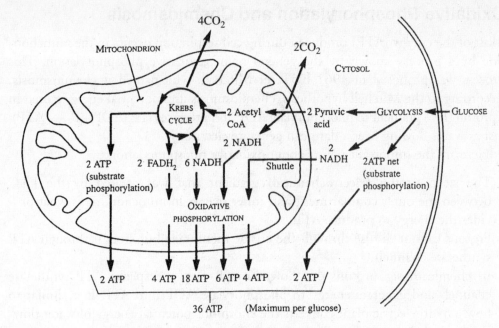

Figure 8.7 Overview of aerobic respiration

MULTIPLE-CHOICE QUESTIONS

1. All of the following are correct about cell respiration EXCEPT

    (A) it is an oxidative process
    (B) all plant and animal cells carry out respiration all the time
    (C) it consists of glycolysis, Krebs cycle, and the electron transport chain
    (D) oxygen molecules are pumped across the cristae membrane to the outer compartment
    (E) most ATP is produced during cell respiration as a result of oxidative phosphorylation

2. Glycolysis

    (A) produces ATP by oxidative phosphorylation
    (B) requires oxygen
    (C) involves an electron transport chain
    (D) is the breakdown of glucose to produce pyruvic acid
    (E) produces carbon dioxide

3. The electron transport chain is located in the

    (A) cytoplasm
    (B) cristae of the mitochondria
    (C) outer compartment of mitochondria
    (D) inner matrix
    (E) Golgi body

4. Each NAD molecule carrying protons to the electron transport chain can produce _____ molecules of ATP.

   (A) 1
   (B) 2
   (C) 3
   (D) 20
   (E) an unlimited number of

5. Which of the following is the most important thing that occurs during cell respiration?

   (A) CO_2 is produced during the Krebs cycle.
   (B) Protons and electrons combine with oxygen to form water.
   (C) Oxygen is released.
   (D) A proton gradient is created.
   (E) ATP is produced.

6. The loss of electrons is known as

   (A) reduction
   (B) oxidation
   (C) redox
   (D) electron transport chain
   (E) lysis

7. The immediate result of the electron transport chain is

   (A) water is produced
   (B) ATP is produced
   (C) lactic acid is produced
   (D) glucose is broken down into 2 pyruvic acid molecules
   (E) a proton gradient is created

8. The role of oxygen in aerobic respiration is

   (A) to transport CO_2
   (B) as the final proton and electron acceptor in the electron transport chain
   (C) to transport electrons in glycolysis
   (D) to provide electrons for the electron transport chain
   (E) most important in the Krebs cycle

9. Which of the following is a waste product of aerobic cell respiration?

   I. Water
   II. Carbon dioxide
   III. ATP

   (A) I only
   (B) II only
   (C) I and II only
   (D) III only
   (E) I, II, and III

Questions 10–15

The three circles represent three major processes in aerobic respiration.

Glucose → (Process *A*) → (Process *B*) → (Process *C*) → $CO_2 + H_2O$

10. Process *A* represents

    (A) glycolysis
    (B) the Krebs cycle
    (C) the electron transport chain
    (D) alcohol fermentation
    (E) lactic acid fermentation

11. Process *A* occurs in

    (A) mitochondria
    (B) chloroplasts
    (C) the nucleus
    (D) the cytoplasm
    (E) different organelles in different cells

12. Process *C* represents

    (A) glycolysis
    (B) the Krebs cycle
    (C) the lectron transport chain
    (D) alcohol fermentation
    (E) lactic acid fermentation

13. Which process produces the most amount of energy?

    (A) *A*
    (B) *B*
    (C) *C*

14. Which is true about process *C*?

    (A) Energy is produced by substrate level phosphorylation.
    (B) It is universal. It is the way all organisms produce energy.
    (C) It involves the membrane structure ATP synthetase.
    (D) It only occurs in animals, not plants.
    (E) It results in the by-product pyruvic acid.

15. Which is true about process *B*?

    (A) It produces the most energy of all the processes.
    (B) It produces energy by oxidative phosphorylation.
    (C) Its production of energy is explained by the Mitchell hypothesis.
    (D) The raw material is pyruvic acid.
    (E) It releases carbon dioxide.

16. Which of the following probably evolved first?

    (A) the Krebs cycle
    (B) oxidative phosphorylation
    (C) glycolysis
    (D) the electron transport chain
    (E) the citric acid cycle

17. The ATP produced during glycolysis and the Krebs cycle is generated by which of the following?

    (A) electron transport chain
    (B) substrate level phosphorylation
    (C) Krebs cycle
    (D) oxidative phosphorylation
    (E) citric acid cycle

18. The breakdown of glucose into pyruvic acid occurs during the process of

    (A) glycolysis
    (B) lactic acid fermentation
    (C) the Krebs cycle
    (D) the electron transport chain
    (E) alcohol fermentation

19. Choose the pair that is correctly matched.

    (A) Krebs cycle—cytoplasm
    (B) Electron transport chain—outer compartment of mitochondria
    (C) Glycolysis—cristae membrane
    (D) Glycolysis—mitochondria
    (E) Electron transport chain—cristae membrane

20. During strenuous exercise, skeletal muscles use up _____ and produce large amounts of _____ , which causes pain and fatigue in the muscle.

    (A) pyruvic acid—carbon dioxide
    (B) carbon dioxide—pyruvic acid
    (C) pyruvic acid—lactic acid
    (D) oxygen—lactic acid
    (E) glycogen—carbon dioxide

EXPLANATION OF ANSWERS

1. **(D)** Protons (H^+), not oxygen, are pumped across the cristae membrane. The role of oxygen is to attract electrons flowing in the electron transport chain. Each oxygen ion combines with 2 electrons and 1 proton to produce 1 molecule of water, a waste product of cell respiration.

2. **(D)** Glycolysis: *glyco* means "sweet" (sugar), *lyse* means "to break apart," and *sis* means "process." Therefore, glycolysis means the process by which sugar is broken down. Pyruvic acid or pyruvate is a 3-carbon molecule and is essentially half a glucose molecule.

3. **(B)** The electron transport chain is a collection of molecules embedded in the cristae membrane of mitochondria. There are thousands of copies of the ETC in every mitochondrion. This greatly enhances the amount of ATP produced.

4. **(C)** NAD delivers electrons to a higher energy level in the ETC than does FAD. Theoretically, each NAD produces 3 ATP molecules, while each FAD produces 2 ATP molecules from oxidative phosphorylation and the electron transport chain.

5. **(E)** Although answers (A), (B), and (D) are all correct about cell respiration, the purpose of the process is to produce energy in the form of ATP. Statement (C) is incorrect. Oxygen is *used* during cell respiration. Oxygen is released during photosynthesis, not during respiration.

6. **(B)** Oxidation is the loss of electrons; reduction is the gain of electrons. Redox refers to a reaction where electrons are both gained and lost. The electron transport chain is a set of redox reactions where electrons are lost and gained by molecules like cytochromes. Lysis is a process of breaking apart. For example, a cell is lysed if it is placed into a hypertonic solution.

7. **(E)** The electron transport chain (ETC) pumps protons across the cristae membrane of mitochondria in order to create a proton gradient. This gradient will power the production of ATP during the process known as oxidative phosphorylation. Water is produced after protons flow through ATP synthetase channels. Lactic acid is produced as a consequence of fermentation. Glucose is broken down into 2 pyruvic acid molecules during glycolysis.

8. **(B)** Oxygen combines with protons (H^+) that have passed through the ATP synthetase channels during oxidative phosphorylation after the electron trans-

port chain. It also exerts a strong pull on electrons passing through the electron transport chain.

$$\text{Oxygen ions} + \text{Protons} + \text{Electrons} \rightarrow \text{Water}$$

Water is the waste product of cell respiration. That is its only role in cell respiration. Oxygen is not a part of glycolysis or the Krebs cycle.

9. **(C)** Water and carbon dioxide are waste products. Water is formed as oxygen combines with protons flowing through the ATP synthetase channels during oxidative phosphorylation of aerobic respiration only after the electron transport chain. Carbon dioxide is released from the Krebs cycle. ATP is a product of cellular respiration, not a waste product.

10. **(A)** Glycolysis is the first step toward the complete breakdown of glucose and the production of ATP. The next step is the conversion of pyruvic acid to acetyl coA, which enters the Krebs cycle. Next in line is the electron transport chain and the creation of the proton gradient that produces the most ATP. Alcohol and lactic acid fermentation are not part of aerobic respiration and produce no ATP.

11. **(D)** Process *A* represents glycolysis. Glycolysis takes place in the cytoplasm of all cells. Process *B* represents the Krebs cycle. Process *C* represents the electron transport chain.

12. **(C)** Process *C* represents the electron transport chain. Glycolysis is the first step toward the complete breakdown of glucose and the production of ATP. The next step is the conversion of pyruvic acid to acetyl coA, which enters the Krebs cycle. Next in line is the electron transport chain and the creation of the proton gradient that produces the most ATP. Alcohol and lactic acid fermentation are not part of aerobic respiration and produce no ATP.

13. **(C)** The electron transport chain involves the creation of the proton gradient that produces the most ATP by oxidative phosphorylation. Alcohol and lactic acid fermentation are not part of aerobic respiration and produce no ATP.

14. **(C)** Process *C* represents the electron transport chain. It uses the energy from the exergonic flow of electrons through the cristae membrane to pump protons across the membrane and create a proton gradient. This gradient will power the phosphorylation of ADP to ATP. This process is carried out by all aerobic organisms, including plants. Pyruvic acid is a product of glycolysis and is required for the Krebs cycle.

15. **(E)** Process *B* is the Krebs cycle. Carbon dioxide is a by-product of the Krebs cycle. Pyruvic acid, the product of glycolysis, does not enter the Krebs cycle directly. It first combines with coenzyme A to produce acetyl coA, which does enter the Krebs cycle. The Krebs cycle produces only a small amount of ATP by what is called substrate level phosphorylation, not by oxidative phosphorylation. The Mitchell hypothesis is an attempt to explain how ATP is produced during oxidative phosphorylation.

16. **(C)** Glycolysis occurs during the anaerobic phase of respiration and evolved on ancient Earth. It was possibly the way the first cells produced energy. The Krebs cycle and the electron transport chain evolved after free oxygen appeared on ancient Earth. Together they make up the aerobic phase of cell respiration.

Oxidative phosphorylation is the process of energy production during the electron transport chain. The citric acid cycle is another name for the Krebs cycle.

17. **(B)** Substrate level phosphorylation produces little ATP. It is the way in which ATP is produced during glycolysis and the Krebs cycle. The main source of ATP is chemiosmosis, which occurs during oxidative phosphorylation of the electron transport chain. The citric acid cycle is another name for the Krebs cycle.

18. **(A)** The breakdown of glucose is the first step in cell respiration. It occurs anaerobically and produces 4 ATP molecules. Lactic acid and alcohol fermentation are also anaerobic but use pyruvic acid produced from glycolysis to produce either lactic acid or alcohol and carbon dioxide, respectively. The Krebs cycle and the electron transport chain occur in the presence of oxygen during aerobic respiration.

19. **(E)** Glycolysis occurs in the cytoplasm. The Krebs cycle occurs in the inner matrix of the mitochondria. The electron transport chain occurs in the cristae membrane.

20. **(D)** Lactic acid fermentation occurs during strenuous exercise when the body cannot keep up with the increased demand for oxygen by skeletal muscles. Pyruvic acid converts to lactic acid that builds up in the muscle, causing fatigue and burning.

HEREDITY

Classical Genetics

CHAPTER 9

- Gregor Mendel
- Basics of probability
- Laws of dominance and segregation
- Monohybrid cross and backcross or testcross
- Law of independent assortment
- Incomplete dominance and codominance
- Multiple alleles and polygenic inheritance
- Sex-linked traits
- Genes and the environment and sex-influenced inheritance
- Karyotype and the pedigree
- Mutations and nondisjunction
- Human inherited disorders

The father of modern genetics is Gregor Mendel, an Austrian monk. In the 1850s, he bred garden peas in order to study patterns of heredity. He collected data from hundreds of plants across many generations and applied statistical analysis to his carefully collected data. Mendel was successful and became famous because he brought the mathematical laws of probability to the study of inheritance.

Mendel's work produced three laws: the law of dominance, the law of segregation, and the law of independent assortment.

BASICS OF PROBABILITY

Probability is the likelihood that a particular event will happen. It cannot predict whether a particular event will actually occur. However, if the sample is large enough, it can predict an average outcome.

If you flip a coin, there is a 50 percent chance that it will come up heads and a 50 percent chance that it will come up tails. The chance of either event happening is 1 out of 2 or $\frac{1}{2}$. If you flip a coin three times in a row, the probability of getting

heads is $\frac{1}{2} \times \frac{1}{2} \times \frac{1}{2}$ or $\frac{1}{8}$. You multiply, in this case, because each flip of the coin

is a separate event. The chance that a couple will have a daughter is 50% or $\frac{1}{2}$. The

chance that they will have 2 daughters is $\frac{1}{2} \times \frac{1}{2}$ or $\frac{1}{4}$. Again, you multiply because each birth is a separate event.

Understanding probability is important to the study of genetics because predicting outcomes is what Punnett squares are all about. What is the chance that two brown-eyed people can give birth to a child with blue eyes? That is probability, and that is what this chapter is all about.

LAW OF DOMINANCE

Mendel's first law is the law of dominance. It states that when two organisms, homozygous (pure) for two opposing traits are crossed, the offspring will be hybrid (carry two different **alleles**) but will exhibit only the dominant trait. The trait that remains hidden is the recessive trait.

Parent (P): TT × tt
                       Pure Tall × Pure Dwarf

	T	T
t	Tt	Tt
t	Tt	Tt

Offspring (F$_1$): Tt
                   All hybrid tall   Law of dominance—
                                          all offspring are tall

LAW OF SEGREGATION

The **law of segregation** states that during the formation of gametes, the two traits carried by each parent separate. Figure 9.1 shows this.

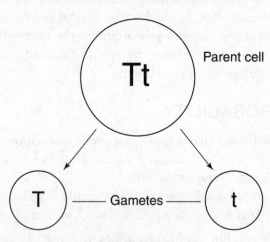

Figure 9.1

The cross that best exemplifies this law is the monohybrid cross (see the next section). In the monohybrid cross, a trait that was not evident in either parent can appear in the F$_1$ generation.

MONOHYBRID CROSS

A **monohybrid cross** is a cross between two organisms that are each hybrid for a single trait, such as $Tt \times Tt$ (T = tall; t = dwarf). The **phenotype** (what the organism looks like) ratios that result from this cross are 3 tall to 1 short, or 75 percent tall plants to 25 percent dwarf plants. The **genotype** (type of genes) ratio is 25 percent homozygous dominant to 50 percent heterozygous to 25 percent homozygous recessive or 1 : 2 : 1. These percentages are always true for a monohybrid cross, and you should memorize them.

F_1: Tt × Tt

	T	t
T	TT	Tt
t	Tt	tt

F_2: TT, Tt, or tt

Monohybrid cross

BACKCROSS OR TESTCROSS

The **testcross** or **backcross** is a way to determine whether an individual plant or animal showing the dominant trait is homozygous dominant (BB) or heterozygous (Bb).

To determine the genotype, the individual of unknown genotype ($B\_$) is crossed with a homozygous recessive individual (bb). The genotype $B\_$ means that one allele is dominant (B) but the other allele is unknown.

If the individual being tested is homozygous dominant (BB), all offspring of the test cross will show the dominant trait and have the hybrid (Bb) genotype. If the individual being tested is actually hybrid (Bb), we can expect that $\frac{1}{2}$ the offspring, or at least one, will show the recessive trait. Therefore, *if any offspring show the recessive trait, the parent of unknown genotype must be hybrid.*

	B	B
b	Bb	Bb
b	Bb	Bb

B = black; b = white

If the parent of unknown genotype is BB, there can be no white offspring.

	B	b
b	Bb	bb
b	Bb	bb

B = black; b = white

> **HELPFUL HINT**
>
> If the phenotype ratio of the offspring is 1 to 1, one parent was hybrid and one parent was pure recessive.

If the parent of unknown genotype is hybrid, there is a 50 percent chance that any offspring will be white.

LAW OF INDEPENDENT ASSORTMENT

The **law of independent assortment** applies when a cross is carried out between two individuals that are hybrid for two traits on separate chromosomes. This law states that during gamete formation, the genes for one trait (such as height, T or t) are not inherited along with the genes for another trait (such as seed color, Y or y). This example will use the following traits: T = tall; t = short; Y = yellow seed; y = green seed.

Figure 9.2 represents a dihybrid individual (*TtYy*) where T will be inherited along with *Y*, while *t* will be inherited along with *y*. The trait for tall (*T*) will be separated from the trait for dwarf (*t*), and the trait for yellow seeds (*Y*) will be separated from the trait for green seeds (*y*).

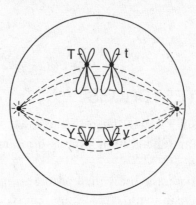

Figure 9.2

The only factor that determines how these alleles are inherited is how the homologous pairs (in this case *TY* and *ty*) happen to line up in metaphase of meiosis I, which is a random event.

During metaphase I, if the homologous pairs happen to line up like this:

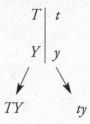

they will produce these gametes.

Instead, if the homologous pairs happen to line up like this:

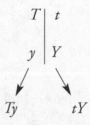

they will produce these gametes.

Now follow a cross that demonstrates the law of independent assortment, a dihybrid cross.

Crossing *Tt Yy* × *Tt Yy* is called a dihybrid cross because it is a cross between individuals that are hybrid for two different traits, in this case, height and seed color. This cross can produce four different types of gametes: *TY*, *Ty*, *tY*, and *ty*. The following figure shows how to set up the Punnett square for this cross.

Gametes of other parent ↓	*TY*	*Ty*	*tY*	*ty*	← Gametes of one parent
TY	*TTYY*	*TTYy*	*TtYY*	*TtYy*	
Ty	*TTYy*	*TTyy*	*TtYy*	*Ttyy*	
tY	*TtYY*	*TtYy*	*ttYY*	*ttYy*	
ty	*TtYy*	*Ttyy*	*ttYy*	*ttyy*	

As you can see, there are many different genotypes possible in the offspring, but you should not pay attention to them. However, the phenotype ratios of the dihybrid cross are significant. The phenotype ratio from a dihybrid cross is 9:3:3:1, 9 tall yellow, 3 tall green, 3 short yellow, and 1 short green. When shown as probabilities, this is:

F_2: 9/16 tall yellow   3/16 tall green   3/16 short yellow   1/16 short green

INCOMPLETE DOMINANCE

Incomplete Dominance is characterized by blending. Here are two examples.

1. A long watermelon (*LL*) crossed with a round watermelon (*RR*) produces all oval watermelon (*RL*).

2. A black animal (*BB*) crossed with a white (*WW*) animal produces all gray (*BW*) animals.

Since neither trait is dominant, the convention for writing the genes uses different capital letters.

A red Japanese four o'clock flower (*RR*) crossed with a white Japanese four o'clock flower (*WW*) produces all pink offspring (*RW*).

Punnett square   *RR* × *WW*

	R	*R*
W	*RW*	*RW*
W	*RW*	*RW*

If two pink four o'clocks are crossed, there is a 25 percent chance that the offspring will be red, a 25 percent chance the offspring will be white, and a 50 percent chance the offspring will be pink.

Punnett square   *RW* × *RW*

	R	*W*
R	*RR*	*RW*
W	*RW*	*WW*

CODOMINANCE

In **codominance**, both traits show. A good example is the *MN* blood group in humans (these are not related to the *ABO* blood group). There are three different blood groups: *M, N,* and *MN*. These groups are based on two distinct molecules located on the surface of the red blood cells. A person can be homozygous for one type of molecule (*MM*), homozygous for the other molecule (*NN*), or be hybrid and have both molecules (*MN*) on the surface of their red blood cells. The *MN* genotype is not intermediate between *M* and *N*, both *M* and *N* traits are expressed, as shown on Figure 9.3.

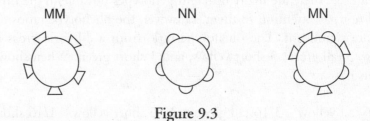

Figure 9.3

MULTIPLE ALLELES

Most genes in a population exist in only two allelic forms. For example, pea plants can be either tall (*T*) or short (*t*). When there are more than two allelic forms of a gene, we refer to that situation as **multiple alleles**. In humans there are four different blood groups: A, B, AB, and O determined by the presence of specific molecules on the surface of the red blood cells. These four different blood types are determined by three alleles, *A, B,* and *O. A* and *B* are codominant and are often written as I^A and I^B. (*I* stands for immunoglobin.) When both alleles are present, they are both expressed, and the person has *AB* blood type. In addition, *O* is a recessive trait and is often written as *i*. A person can have any one of the six blood genotypes shown in Table 9.1.

TABLE 9.1

Human Blood Types and Genotypes		
Blood Type	**Genotype**	
A	Homozygous *A*:	*AA*
A	Hybrid *A*:	*Ai*
B	Homozygous *B*:	*BB*
B	Hybrid *B*:	*Bi*
AB	Heterozygous:	*AB*
O	Homozygous *O*:	*ii*

POLYGENIC INHERITANCE

Many characteristics such as skin color, hair color, and height result from a blending of several separate genes that vary along a continuum. They are controlled by several genes and are called **polygenic**. Two parents who are short carry more genes for shortness than for tallness. However, they can have a child who inherits mostly genes for tallness from both parents and who will be taller than his/her parents. This wide variation in genotypes always results in a bell-shaped curve in an entire population, as seen in Figure 9.4 which shows the distribution of skin pigmentation across a population.

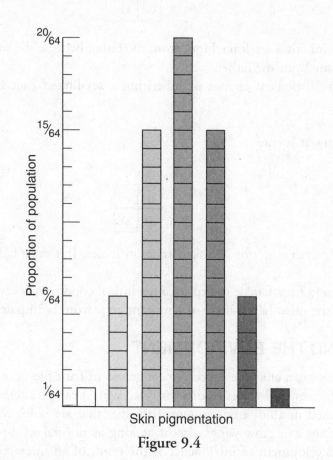

Figure 9.4

SEX-LINKED GENES

Traits carried on the X chromosome are called **sex-linked**. Females (XX) inherit two copies of sex-linked genes. If a sex-linked trait is due to a recessive mutation, a female will express the phenotype only if she carries two mutated genes (X–X–). If she carries only one mutated X-linked gene, she will be a carrier (X–X). If a sex-linked trait is due to a **dominant mutation**, a female will express the phenotype with only one mutated gene (X–X). Males (XY) inherit only one X-linked gene. As a result, if the male inherits a mutated X-linked gene (X–Y), he will express the gene. Recessive sex-linked traits are much more common than dominant sex-linked traits; so males suffer with sex-linked conditions more often than females do.

Here are some important facts about sex-linked traits.

- Common examples of recessive sex-linked traits are color blindness and **hemophilia.**
- All daughters of affected fathers are carriers.

Punnett square

	X–	Y
X	X–X	XY
X	X–X	XY

- Sons cannot inherit a sex-linked trait from the father because the son inherits the Y chromosome from the father.
- A son has a 50 percent chance of inheriting a sex-linked trait from a carrier mother.

REMEMBER

In sex-linked inheritance, the father passes the trait to his daughters only.

Punnett square

	X	Y
X–	X–X	X–Y
X	XX	XY

- There is no carrier state for X-linked traits in males. If a male has the gene, he will express it.
- It is uncommon for a female to express a sex-linked condition because in order to be affected, she must have inherited a mutant gene from both parents.

GENES AND THE ENVIRONMENT

The environment can alter the expression of genes. In fruit flies, the expression of the mutation for vestigial wings (short, shriveled wings) can be altered by temperature. When raised in a hot environment, fruit flies that are homozygous recessive for vestigial wings can grow wings almost as long as normal wild-type wings. In humans, the development of intelligence is the result of an interaction of genetic predisposition and the environment, or "nurture versus nature."

SEX-INFLUENCED INHERITANCE

Inheritance can be influenced by the sex of the individual carrying the traits. An example can be seen in male pattern baldness in humans, where hair is very thin on top of the head. This is not a sex-linked trait but rather a **sex-influenced trait.** Males and females express the gene for pattern baldness differently, as seen in Table 9.2.

TABLE 9.2

Sex Influence on Pattern Baldness

Genotype	Phenotype	
	Female	**Male**
BB	Bald	Bald
Bb	Not bald	Bald
bb	Not bald	Not bald

B = bald; *b* = not bald

KARYOTYPE

A karyotype is a labaoratory procedure that analyzes the size, shape, and number of chromosomes. Specialists prepare and photograph chromosomes during metaphase of mitosis when they are fully condensed. Of the 46 human chromosomes, there are 44 (22 pairs of) **autosomes** and 2 sex chromosomes. Figure 9.5 shows a karyotype of a normal male. Notice the X and Y sex chromosomes. Females in contrast, have XX.

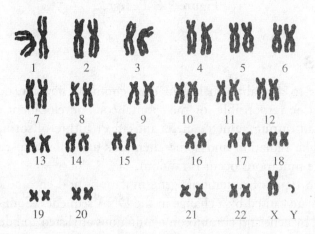

Figure 9.5 Karyotype of normal male

THE PEDIGREE

A **pedigree** is a family tree that indicates the phenotype of one trait being studied for every member of a family. Geneticists use the pedigree to determine how a particular trait is inherited. By convention, females are represented by a circle and males by a square. The carrier state is not always shown. If it is, it is sometimes represented by a half-shaded-in shape. A shape is completely shaded in if a person exhibits the trait.

The pedigree in Figure 9.6 shows three generations of deafness. Try to determine the pattern of inheritance. First, eliminate any possibilities. The trait is not dominant (either sex-linked or autosomal). In order for a child to have the condition, she or he would have had to have received one mutant gene from one afflicted parent, and nowhere is that the case. (All afflicted children have unaffected parents.) Also, the trait is not sex-linked recessive because in order for F_3 generation daughter #1 to have the condition, she would have had to inherit two mutant traits (X–X–), one from each parent. Yet her father does not have the condition. Since you have eliminated all the possibilities it could NOT be, you must conclude that the trait for deafness must be autosomal recessive.

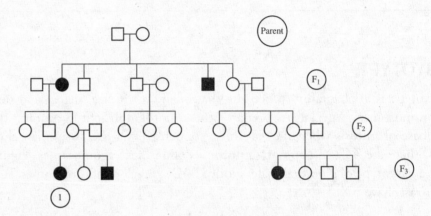

Figure 9.6 Pedigree

MUTATIONS

Mutations refer to any abnormality in the genome. They can occur in somatic (body) cells and be responsible for the spontaneous development of cancer. They can occur instead during gametogenesis and affect future offspring. Even though certain things like radiation and some chemicals are known to cause mutations, when and where mutations occur is random.

There are two types of mutations, gene mutations and chromosome mutations. Gene mutations are caused by a change in the DNA sequence. Some human genetic disorders caused by gene and chromosome mutations are listed and described in Table 9.3. The nature of gene mutations at the DNA level is discussed in the next chapter.

Chromosome mutations can be observed under a light microscope. A chromosome may sustain a deletion or an addition, or a cell may have an entirely extra chromosome. This last mutation is due to nondisjunction.

NONDISJUNCTION

Nondisjunction is an error that sometimes happens during meiosis in which homologous chromosomes fail to separate, as they should. This is illustrated in Figure 9.7. When this happens, one gamete receives two homologues, while the other gamete receives none. The remaining chromosomes may be unaffected and normal. If either of these aberrant gametes unites with a normal gamete during fertilization, the resulting zygote will have an abnormal number of chromosomes.

Figure 9.7 Nondisjunction

Any abnormal chromosome condition is known as **aneuploidy**. If a chromosome is present in triplicate, the condition is known as **trisomy**. People with Down syndrome have three #21 chromosomes. The condition is referred to as trisomy-21.

An organism in which the cells have an extra set of chromosomes is referred to as triploid (3*n*). The cells of the endosperm or cotyledon of a seed are triploid. An organism with more than 3 sets of chromosomes is referred to as **polyploid**. Scientists breed plants to be polyploid because they will produce abnormally large flowers and fruit.

HUMAN INHERITED DISORDERS

Human genetic defects can be caused by either a gene or a chromosome mutation. (The cause of a gene mutation is discussed in the next chapter.) Gene mutations are not visible under a microscope, but chromosomal mutations can be assessed by doing a procedure called a karyotype. A karyotype is carried out on cells from a developing fetus to scan for chromosomal abnormalities such as trisomy-21, Down syndrome.

Chromosomal aberrations include:

1. **Deletion.** A fragment lacking a centromere is lost during cell division.

2. **Inversion.** A chromosomal fragment reattaches to its original chromosome but in the reverse orientation.

3. **Translocation.** A fragment of a chromosome becomes attached to a non-homologous chromosome.

4. **Polyploidy.** A cell or organism has extra sets of chromosomes.

5. **Nondisjunction.** Homologous chromosomes fail to separate during meiosis.

TABLE 9.3

Gene and Chromosome Mutations

Genetic Disorder	Pattern of Inheritance	Description
Phenylketonuria (PKU)	Autosomal recessive	Inability to break down the amino acid phenylalanine. Requires elimination of phenylalanine from diet; otherwise serious mental retardation will result.
Cystic fibrosis	Autosomal recessive	The most common lethal genetic disease in the U.S. 1 out of 25 Caucasians is a carrier. Characterized by buildup of extracellular fluid in the lungs, digestive tract, etc.
Tay-Sachs disease	Autosomal recessive	Onset is early in life and is caused by lack of the enzyme needed to break down lipids necessary for normal brain function. It is common in Ashkenazi Jews and results in seizures, blindness, and early death.
Huntington's disease	Autosomal dominant	A degenerate disease of the nervous system resulting in certain and early death. Onset is usually in middle age.
Hemophilia	Sex-linked recessive	Caused by the absence of one or more proteins necessary for normal blood clotting.
Color blindness	Sex-linked recessive	Red-green color blindness is rarely more than an inconvenience.

Chromosomal Disorder	Pattern of Inheritance	Description
Down syndrome	47 chromosomes with trisomy-21	Characteristic facial features, mental retardation, prone to developing Alzheimer's and leukemia.
Klinefelter's syndrome	XXY 47 chromosomes	Have male genitals, but the testes are abnormally small and the men are sterile.

MULTIPLE-CHOICE QUESTIONS

1. A hybrid red flower (*Rr*) is crossed with a white flower. What percentage of the offspring will be red?

   (A) 0%
   (B) 25%
   (C) 50%
   (D) 75%
   (E) 100%

2. A red flower is crossed with a white flower, and all the offspring are pink. What law of inheritance does this follow?

   (A) Dominance
   (B) Segregation
   (C) Incomplete dominance
   (D) Law of independent assortment
   (E) Codominance

3. Which of the following crosses best demonstrates the law of segregation?

   (A) *AA* × *aa*
   (B) *aa* × *aa*
   (C) *Aa* × *aa*
   (D) *Aa* × *Aa*
   (E) *AA* × *AA*

4. If two pink Japanese four o'clock flowers (*RW*) are crossed, what is the chance that an offspring will be pink?

   (A) 0%
   (B) 25%
   (C) 50%
   (D) 75%
   (E) 100%

5. A couple has two children. One child has blood type A, and the other child has blood type O. What are all the possible blood types of the parents?

   (A) Either both have type A, or one has type A and the other has type O, or one has A and the other has type B.
   (B) There is only one possibility. Both parents have type A blood.
   (C) There is only one possibility. Both parents have type O blood.
   (D) There is only one possibility. One parent has type A, and the other has type O.
   (E) It is not possible to determine from the information given.

6. Which is true about blood type?

   (A) Type A is dominant over type B.
   (B) Type B is dominant over type A.
   (C) Type O is dominant over types A and B.
   (D) Type AB is dominant over both A and B.
   (E) Types A and B are dominant over type O.

7. A man with hemophilia marries a woman who has normal hemoglobin and is not a carrier. Which of the following is true?

   (A) None of their children will have the disease nor will they be carriers.
   (B) All the boys will have the disease.
   (C) All the girls will have the disease.
   (D) All the boys will be carriers.
   (E) All the girls will be carriers.

8. If a man is color-blind and his wife is a carrier, which of the following is true?

   (A) Each girl has a 50% chance of being color-blind.
   (B) No girl will be color-blind because girls cannot be color-blind.
   (C) All the boys will be carriers.
   (D) All the girls will be carriers.
   (E) All children will be either a carrier or color-blind.

Questions 9–14

Refer to the list below of inherited diseases.

   (A) Phenylketonuria
   (B) Cystic fibrosis
   (C) Huntington's disease
   (D) Hemophilia
   (E) Down syndrome

9. Results from a mutation of chromosomes

10. Characterized by a buildup of fluid in the lungs

11. If no dietary change is instituted at birth, mental retardation will result

12. Autosomal dominant disease of the nervous system that results in death

13. Sex-linked recessive disorder

14. Onset is usually in middle age

15. A man with Klinefelter's syndrome has 47 chromosomes and has three sex chromosomes, XXY. This abnormality came about as a result of

    (A) crossing-over
    (B) deletion of a chromosome
    (C) addition of a chromosome
    (D) nondisjunction
    (E) mutation in a gene

16. A testcross is done to

    (A) determine if crossover has occurred
    (B) prevent sex-linked traits from being passed on
    (C) determine if a person has Huntington's disease
    (D) prevent crossover from occurring
    (E) determine if an organism with the dominant phenotype is homozygous dominant or hybrid

17. A farmer planted 1,000 seeds of corn. The offspring were 544 tall with yellow seeds, 188 tall with green seeds, 183 short with yellow seeds, and 64 short with green seeds. What were the genotypes of the parents?

    (A) *TTYY × ttyy*
    (B) *TtYy × ttyy*
    (C) *ttyy × ttyy*
    (D) *ttyy × TtYy*
    (E) *TtYy × TtYy*

18. The genotype ratio from a monohybrid cross is

    (A) 1:1:1 homozygous dominant : hybrid : homozygous recessive
    (B) 3:1 homozygous dominant : hybrid
    (C) 1:2:1 homozygous dominant : hybrid : homozygous recessive
    (D) 3:1 hybrid : homozygous recessive
    (E) cannot be stated universally

Questions 19–20

Refer to this hypothetical pedigree of an inherited trait in humans.

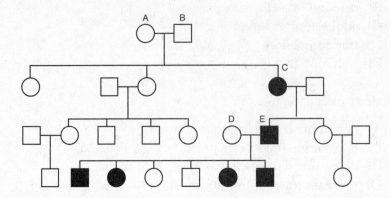

19. Which is true of the shaded-in trait?

    (A) The trait is autosomal recessive.
    (B) The trait is autosomal dominant.
    (C) The trait is sex-linked dominant.
    (D) The trait is sex-linked recessive.
    (E) The trait is inherited as incomplete dominance.

20. What is the most likely genotype for person C?

    (A) *AA*
    (B) *Aa*
    (C) *aa*
    (D) *X–X–*
    (E) *X–X*

EXPLANATION OF ANSWERS

1. **(C)** The fact that the flower is hybrid red means that red is dominant. The fact that there is a white flower means that white is recessive. So the cross is between *Rr* and *rr*. There is a 50 percent chance that the offspring will be red and a 50 percent chance that it will be white.

2. **(C)** The pattern of inheritance is incomplete dominance or blending because red and white makes pink. Other examples would be an animal with black fur crossed with an animal with white fur yielding gray offspring or a round watermelon crossed with a long watermelon yielding oval offspring.

3. **(D)** The law of segregation states that alleles separate during gamete formation. This is best proved by answer (D) because both parents express the dominant trait, *A*, but some offspring would show the recessive trait. This could not happen if it were not for homologous chromosomes separating during meiosis. In answers (A) and (C), you see both traits in the parents and both traits in the offspring. There is nothing odd about that. In answers (B) and (E),

the parents appear with one phenotype and the offspring have the same phenotype. Again, that does not relate to segregation.

4. **(C)** The pattern of inheritance here is blending or incomplete dominance. If you cross $RW \times RW$, there will be a 25 percent chance that the offspring will be red, 50 percent chance of pink, and 25 percent chance of white.

5. **(A)** In order for a child to have type O, both parents must contribute one O allele. In this case, both parents could be hybrid A (Ai), or one parent could have type A (Ai) and the other could have type O (ii), or one parent A (Ai) and the other type B (Bi).

6. **(E)** Blood types A and B are dominant over type O. Both A and B are codominant with respect to each other. If someone has both A and B alleles, they are both present on the surface of the red blood cells.

7. **(E)** When doing Punnett squares for sex-linked traits, consider female and male offspring separately. If the question asks, "What is the chance of having a girl with the condition?" you are limited to 0 percent, 50 percent, or 100 percent because there are only two boxes to consider for each sex. In addition, males cannot be carriers of sex-linked traits. Either they have the condition or they do not. Only females can be carriers. In this cross, all the females will be carriers and the boys will be normal. Here is an important rule: all daughters of fathers with a sex-linked trait will be carriers.

The cross looks like this:

	X–	Y
X	X–X	XY
X	X–X	XY

8. **(A)** Here is the cross:

	X–	Y
X–	X–X–	X–Y
X	X–X	XY

When doing Punnett squares for sex-linked traits, consider female and male offspring separately. If the question asks, "What is the chance of having a girl with the condition?" you are limited to 0 percent, 50 percent, or 100 percent because there are only two boxes to consider for each sex. In this cross, the chance of a girl having the condition is 50 percent, while the chance of having a normal girl is also 50 percent. The chance of having a boy who is normal is 50 percent, while a boy with the condition is 50 percent.

9. **(E)** Down syndrome results from having an extra chromosome 21, which occurs as a result of nondisjunction during gamete formation. A person with Down syndrome suffers from mental retardation and several other physical problems.

10. **(B)** Cystic fibrosis is the most common lethal genetic disease in the U.S. It is inherited as an autosomal recessive gene.

11. **(A)** Every newborn in the U.S. is tested for phenylketonuria (PKU). A baby with the disease cannot process the amino acid phenylalanine. Therefore, phenylalanine must be eliminated from the diet. Otherwise, poisons will build up in the brain and mental retardation will result.

12. **(C)** Huntington's disease is a degenerative disease of the nervous system that results in death. It is one of the few inherited conditions that is dominant. The onset is in middle age, usually after a person has already had children. A child of a person with Huntington's has a 50 percent chance of inheriting the disease.

13. **(D)** Hemophilia is a disease where a person's blood does not clot normally. It is inherited as a sex-linked recessive trait.

14. **(C)** The onset of Huntington's disease is usually in middle age. This is particularly unfortunate because the person might have already had children and passed the disease on to his or her offspring. Huntington's disease is not treatable.

15. **(D)** This condition, like any condition that results in an abnormal number of chromosomes, is caused by nondisjunction, the failure of homologous chromosomes to separate during meiosis.

16. **(E)** A testcross is an actual breeding of an animal or plant to determine whether the organism is homozygous dominant (*BB*) or heterozygous (*Bb*). Both conditions would appear identical. The individual of unknown genotype is crossed with a homozygous recessive individual. If any offspring show the recessive trait, then the parent showing the dominant trait must have been hybrid.

17. **(E)** The ratios of the offspring are closest to 9:3:3:1, and therefore this is a dihybrid cross. Each parent plant is hybrid for both traits. The fact that the total number of offspring does not add up to 1,000 is of no concern. You do not usually expect 100 percent of seedlings to sprout.

18. **(C)** You must memorize the phenotypic and genotypic ratios from a monohybrid cross. The genotype ratios are 1 homozygous dominant:2 hybrid:1 homozygous recessive. The phenotype ratios 3 dominant:1 recessive.

19. **(B)** When analyzing a pedigree, always begin by eliminating choices first. Look at the possibility of a sex-linked trait first. The trait appears for the first time in female C. She gave the trait to her son but not to her daughter, so you might expect it to be sex-linked. However, look at the 6 children of D and E; 4 out of 6 children, both boys and girls inherited the disease. If it were sex-linked, moms would pass the trait to sons, and dads would pass the trait to daughters. That is not the case. So the trait is not sex-linked. Therefore, it must be autosomal. Now see if it could be dominant or recessive. We can only take a guess as to what is most likely, but it looks like a mutation occurred in a gamete of parent C. It seems to be a rare mutation because it is not in any other family members except in the children of D and E. Because it is so common in that family, it is most likely an autosomal dominant trait.

20. **(B)** Since the trait is inherited as autosomal dominant, person C would have to be *Aa*.

Molecular Genetics

- The search for inheritable material
- Structure of DNA and RNA
- DNA replication
- How DNA makes protein
- Gene regulation
- Mutation

- Basics about the human genome
- Genetic engineering and recombinant DNA
- Restriction enzymes, gel electrophoresis, and PCR

Today, everyone knows that deoxyribonucleic acid (DNA) is the molecule of heredity. However, that fact has not always been widely accepted. Until the early 1950s, many scientists believed that proteins were the molecules that make up genes and constitute inherited material. The work of many brilliant scientists has transformed our knowledge of the structure and function of the DNA molecule and led to the acceptance of DNA as the molecule responsible for heredity.

THE SEARCH FOR INHERITABLE MATERIAL

Griffith (1927) discovered the natural phenomenon known as bacterial transformation, which is the ability of bacteria to alter their genetic makeup by uptaking foreign DNA from another bacterial cell and incorporating it into their own. He worked with different strains of the bacterium, *Diplococcus pneumoniae*, which cause pneumonia.

Avery, MacLeod, and McCarty (1944) published their classic findings that the molecule that Griffith's bacteria were transferring was, in fact, DNA. They provided direct experimental evidence that DNA is the genetic material.

Hershey and Chase (1952) proved that DNA, not proteins, is the molecule of inheritance when they tagged bacteriophages with the radioactive isotopes ^{32}P and ^{35}S. The ^{32}P labeled the DNA of the phage viruses, while the ^{35}S labeled the protein coat of the phage viruses. Hershey and Chase found that when bacteria were infected with phage viruses, ^{32}P from the virus entered the bacterium and produced thousands of progeny. However, no ^{35}S entered the bacterium.

Rosalind Franklin (1950–53), continuing the work begun by Maurice Wilkins, carried out the X-ray crystallography analysis of DNA that showed DNA to be a helix. Her work was critical to Watson and Crick in developing their now-famous model of DNA.

> **INTERESTING FACT**
>
> Rosalind Franklin did not share the Nobel Prize with Watson and Crick because she died before it was awarded, and it is not given posthumously.

Watson and Crick received the Nobel Prize in 1962 for correctly describing the structure of DNA as a double helix.

Meselson and Stahl (1953) proved Watson and Crick's hypothesis that DNA replicates in a semiconservative fashion. See Figure 10.1. They cultured bacteria in a medium containing heavy nitrogen (^{15}N) and then moved them to a medium containing light nitrogen (^{14}N), allowing the bacteria to replicate and divide once. The new bacterial DNA contained DNA consisting of one heavy strand and one light strand, thus proving Watson and Crick's theory.

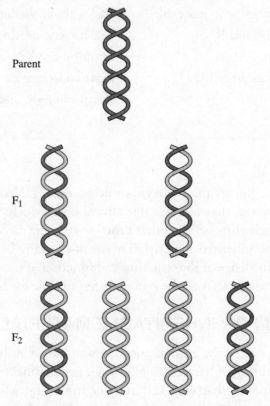

Figure 10.1 Semi-conservative replication

Structure of Deoxyribonucleic Acid (DNA)

Figure 10.2 shows the DNA molecule.

- DNA is a double helix shaped like a twisted ladder.
- DNA consists of two strands running in opposite directions from each other.
- It is a polymer made of repeating units called nucleotides.
- Each nucleotide consists of a 5-carbon sugar (deoxyribose), a phosphate molecule, and a nitrogenous base.
- Each nucleotide contains one of the four possible nitrogenous bases: adenine (*A*), thymine (*T*), cytosine (*C*), and guanine (*G*).
- The nitrogenous bases of opposite chains are paired to one another by **hydrogen bonds**.
- A bonds with T; C bonds with G.

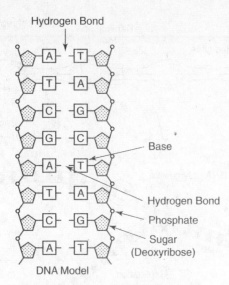

Figure 10.2 DNA molecule

Structure of Ribonucleic Acid (RNA)

- RNA is a single-stranded helix.
- It is a polymer made of repeating units called nucleotides.
- Each nucleotide consists of a 5-carbon sugar (ribose), a phosphate, and a **nitrogenous base**.
- Each nucleotide contains one of the four possible nitrogenous bases: **adenine** (*A*), **uracil** (*U*), **cytosine** (*C*), and **guanine** (*G*).
- There is no thymine in RNA. Uracil replaces thymine.
- There are three types of RNA: mRNA (messenger RNA), tRNA (transfer RNA), and rRNA (ribosomal RNA).

1. **Messenger RNA (mRNA).** It carries messages directly from DNA in the nucleus to the cytoplasm during the making of protein. The triplet nucleotides of mRNA (such as AAC or UUU) are called **codons.**

2. **Transfer RNA (tRNA).** It is shaped like a cloverleaf and carries amino acids to the mRNA at the ribosome in order to form a polypeptide. The triplet nucleotides of tRNA are complementary to the codons of mRNA and are called **anticodons.**

3. **Ribosomal RNA (rRNA)** is structural. Along with proteins, it makes up the ribosome.

DNA REPLICATION IN EUKARYOTES

DNA replication is the making of an exact replica of DNA. The two new molecules of DNA that are produced each consist of one old strand and one new strand. This is called **semiconservative replication** as proved by Meselson and Stahl. This is shown in Figure 10.3.

TIP

DNA replication occurs during interphase of the cell cycle.

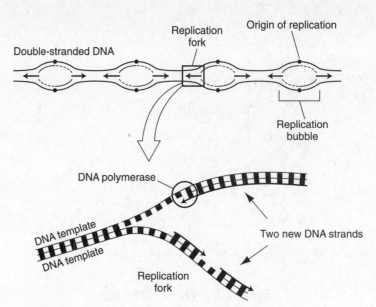

Figure 10.3 **DNA replication in eukaryotes**

- Replication occurs during interphase in the life cycle of a cell.
- DNA polymerase catalyzes the replication of the new DNA.
- DNA polymerase also proofreads each new DNA strand, fixing errors and minimizing the occurrence of mutations.
- DNA unzips at the hydrogen bonds that connect the two strands of the double helix.
- Each strand of DNA serves as a template for the new strand according to the base-pairing rules: A with T and C with G.
- If a strand of DNA to be copied is AAATCGGAC,
    then the new strand is TTTAGCCTG.
- Each time the DNA replicates, some nucleotides from the ends of the chromosomes are lost. To protect against the possible loss of genes at the ends of the chromosomes, some eukaryotic cells have special nonsense nucleotide sequences at the ends of chromosomes that repeat thousands of times. These protective ends of the chromosomes are called **telomeres.**

HOW DNA MAKES PROTEIN

The process whereby DNA makes proteins has been worked out in great detail. There are three main steps: **transcription**, **RNA processing**, and **translation**.

Transcription

Transcription is the process by which DNA makes RNA. It is facilitated by RNA polymerase and occurs in the nucleus. The triplet code in DNA is transcribed into a codon sequence in messenger RNA (mRNA), following the base-pairing rules: A with U and C with G. Remember, there is no thymine in RNA. Uracil replaces thymine.

If the sequence in DNA triplets is: AAA TAA CCG GAC

Then the complementary sequence of
codons in mRNA is: UUU AUU GGC CUG

RNA Processing

After transcription but before the newly formed strand of RNA is shipped out of
the nucleus to the ribosome, this **initial transcript** is processed or edited by a series
of enzymes. The enzymes remove pieces of RNA that do not code for any protein.
These noncoding regions that are removed are called **introns** (intervening
sequences). The remaining portions, **exons** (expressed sequences or coding regions),
are pieced back together to form the **final transcript**. As a result of this processing,
the mRNA that leaves the nucleus is a great deal shorter than the piece that was initially transcribed.

Translation of mRNA into Protein

Translation is the process by which the mRNA sequence is converted into an amino
acid sequence.

- Translation occurs at the ribosome.
- Amino acids present in the cytoplasm are carried by tRNA molecules to the
  codons of the mRNA strand at the ribosome according to the base-pairing rules
  (A with U and C with G).
- Some tRNA molecules can bind to two or more different codons. For example,
  codons UCU, UCC, UCA, and UCG all code for a single amino acid, serine.
  Figure 10.4 shows the translation of the RNA code into an amino acid sequence.

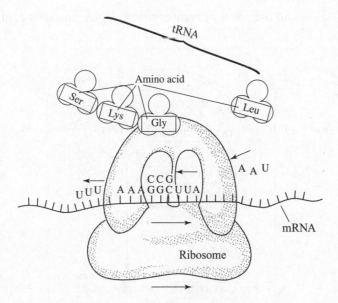

Figure 10.4 Translation

Table 10.1 shows a sample of the genetic code. It lists several mRNA codons and
the amino acids that are translated from that code.

TABLE 10.1

Codons and Amino Acids

Codon Sequence	Codons and Amino Acids
UUU	Phenylalanine
AGU	Serine
GUU	Valine
GGU	Glycine
CUU	Leucine
AUG	Methionine

By using the code in Table 10.1, determine what amino acid sequence will form if the DNA sequence is:

TACAAACAAGAATCA

First, transcribe the DNA triplets into mRNA codons:

AUGUUUGUUCUUAGU

Then, use the genetic code to translate the codon sequence to an amino acid sequence:

Methionine-Phenylalanine-Valine-Leucine-Serine

Figure 10.5 shows an overview of transcription, RNA processing, and translation.

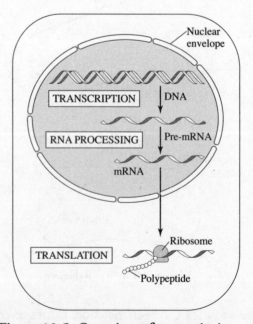

**Figure 10.5 Overview of transcription,
RNA processing, and translation**

GENE REGULATION

Every cell does not constantly synthesize every polypeptide it has the ability to make. For example, cells in the pancreas are not always producing tons of insulin because it is not always needed in that quantity. That means that every gene in a cell is not turned on all the time. How does a cell know when to turn on a gene or when to turn it off? This is actually a very complex process in humans and one that is not understood well. However, a simple model for gene regulation can be found in bacteria in a region of DNA called the operon.

The operon is actually a cluster of functional genes plus the switches that turn them on and off. There are two types of operons. One is the Lac or inducible operon, which is normally turned off unless it is actively induced or triggered to turn on by something in the environment. The other is the repressible operon, which is always turned on unless it is actively turned off because it is temporarily not needed.

You need to know some of the parts of the operon. The two most relevant ones are the promoter and the operator. The promoter is the binding site of RNA polymerase. RNA polymerase must always bind to DNA before transcription can take place. The other important region is the operator. This is the binding site for the repressor, which turns off the Lac operon. Another part of the operon, with a very funny name, is the TATA box (named for its sequences of alternating adenine and thymine), which helps RNA polymerase bind to the promoter.

MUTATIONS

Mutations are changes in genetic material. They occur spontaneously and at random and can be caused by mutagenic agents, including toxic chemicals and radiation. Mutations are the raw material for natural selection.

Gene Mutations

Several types of gene mutations can occur: point mutations, insertions, and deletions. Both types can have deleterious effects on the organism.

POINT MUTATION

The simplest mutation is a **point mutation**. This is a **base-pair substitution**, where one nucleotide converts to another. Here is an example of a change in an English sentence analogous to a point mutation in DNA:

Normal		Point Mutation
↓		↓
THE FAT CAT SAW THE **DOG**	*becomes*	THE FAT CAT SAW THE **HOG**.

The inherited genetic disorder sickle cell anemia results from a point mutation, like the one shown above, in the gene that codes for hemoglobin. The abnormal hemoglobin causes red blood cells to sickle when available oxygen is low. When red blood cells sickle, a variety of tissues may be deprived of oxygen and suffer severe and permanent damage.

It is possible, however, that a point mutation could result in a beneficial change for an organism or in no change in the proteins produced. Table 10.2 shows one example where a point mutation in DNA would result in no change in the amino acid sequence.

TABLE 10.2

Nonharmful Point Mutation

DNA	mRNA	Amino Acid Produced
AAA	UUU	Phenylalanine
AAG	UUC	Phenylalanine
Mutation ↑	Mutation ↑	No change in the amino acid

INSERTION OR DELETION

A second type of gene mutation results from a single nucleotide **insertion** or **deletion**. To continue the three-letter word analogy, a deletion is the loss of one letter and an insertion is the addition of a letter into the DNA sentence. Both mutations result in a **frameshift** because the entire reading frame is altered.

Deletion of the Letter E

THE FAT CAT SAW THE DOG → THF ATC ATS AWT HED OG

Insertion of the Letter T

THE FAT CAT SAW THE DOG → THE FTA TCA TSA WTH EDO G

Depending on where it occurs, a frameshift can have disastrous results. It can cause the formation of an altered polypeptide or no polypeptide at all.

Chromosome Mutations

Chromosome mutations are alterations in chromosome number or structure and are visible under a microscope. **Aneuploidy** is the term applied to having any abnormal number of chromosomes. One common example is Down syndrome, known officially as **trisomy-21**, where a person is born with an extra chromosome 21. This is seen in the pedigree shown in Figure 10.6.

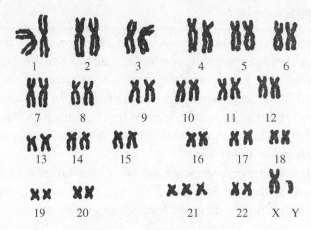

Figure 10.6 Trisomy-21

Having entire extra sets of chromosomes, such as *2n* or *4n*, is known as **polyploidy** and is normal in plants. Polyploidy is responsible for unusually large and brilliantly colored flowers.

Aneuploidy and polyploidy both result from **nondisjunction**, where homologous pairs fail to separate during meiosis.

THE HUMAN GENOME

The human **genome** (an organism's genetic material) consists of 3 billion base pairs of DNA and about 30,000 genes. Surprisingly, 97 percent of our DNA does NOT code for protein product and has often been called **junk**. Of the noncoding DNA, some are regulatory sequences that control gene expression. Some are introns that interrupt genes. However, most of the DNA consists of repetitive sequences that never get transcribed. Many of these tandem repeats consists of short sequences repeated as much as 10 million times. Scientists do not understand very much about junk DNA.

GENETIC ENGINEERING AND RECOMBINANT DNA

Recombinant DNA means taking DNA from two sources and combining them in one cell. The branch of science that uses recombinant DNA techniques for practical purposes is called **genetic engineering** or biotechnology. One of the most important areas of study in genetic engineering is gene therapy. Scientists are trying to learn how to insert functioning genes into cells to replace nonfunctioning ones. If they are successful, it would mean an end to genetic diseases like cystic fibrosis and sickle cell anemia.

Restriction Enzymes

Restriction enzymes are an important tool for scientists working with DNA. They cut DNA at specific **recognition sequences** or **sites**, such as GAATTC, and are sometimes referred to as molecular scissors. The pieces of DNA that result from the cuts made by restriction enzymes are called **restriction fragments**. Hundreds of different restriction enzymes have been isolated from bacteria.

Gel Electrophoresis

Gel electrophoresis separates large molecules of DNA on the basis of their rate of movement through an agarose gel in an electric field. The smaller the molecule, the faster it runs through the gel. If necessary, the concentration of the agarose gel can be changed to provide a better separation of the tiny DNA fragments.

In order to run DNA through a gel, it must first be cut up by restriction enzymes into pieces small enough to migrate through the gel. Once separated on a gel, the DNA can be analyzed in many ways.

Figure 10.7 shows an electrophoresis gel with four samples of DNA that were previously cut with restriction enzymes. Each sample is running its own lane. The shorter pieces of DNA run farther and faster through the gel.

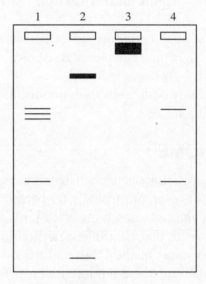

Figure 10.7

Lane 1 has four bands of DNA, three larger pieces and one short piece. Lane 2 contains two pieces of DNA, one large and one tiny. Lane 3 contains one very large and uncut piece of DNA. Lane 4 contains two pieces of DNA.

Polymerase Chain Reaction

Devised in 1985, **polymerase chain reaction (PCR)** is a cell-free, automated technique by which a piece of DNA can be rapidly copied or amplified. Billions of copies of a fragment of DNA can be produced in a few hours. Once the DNA is amplified, these copies can be studied or used in a comparison with other DNA samples.

MULTIPLE-CHOICE QUESTIONS

1. All of the following about the structure of DNA are correct EXCEPT

   (A) DNA is a polymer
   (B) DNA contains deoxyribose
   (C) the two strands are connected by hydrogen bonding
   (D) adenine bonds to guanine
   (E) nucleotides consist of a sugar, phosphate, and nitrogenous base

2. If guanine makes up 28 percent of the nucleotides in a sample of DNA from an organism, then thymine would make up _____ percent of the nucleotides.

   (A) 0
   (B) 22
   (C) 28
   (D) 44
   (E) 56

3. All of the following statements about RNA are correct EXCEPT

   (A) there are three kinds of RNA: mRNA, rRNA, and tRNA
   (B) RNA stands for ribonucleic acid
   (C) RNA is single stranded
   (D) RNA contains the bases C, G, A, and T
   (E) RNA contains the sugar ribose

4. DNA can be cut into small pieces with

   (A) introns
   (B) exons
   (C) PCR
   (D) restriction enzymes
   (E) RNA

5. If the sequence of a strand of DNA is AAATAACCGGGGATC, then the sequence on the corresponding mRNA strand is

   (A) AAAUAACCGGGGATC
   (B) TTTATTGGCCCCTAG
   (C) UUUTUUGGCCCCUAG
   (D) TTTUTTGGCCCCTAG
   (E) UUUAUUGGCCCCUAG

6. All of the following are correct about RNA processing EXCEPT

   (A) it is a rare occurrence and results in mutations
   (B) introns are removed
   (C) it occurs after transcription
   (D) the initial transcript is longer than the final transcript
   (E) it occurs in the nucleus

Questions 7–13

Choose from the terms below.

      (A) Translation
      (B) Transcription
      (C) Replication
      (D) RNA processing
      (E) None of the above

7. Synthesis of RNA

8. Requires RNA polymerase

9. Synthesis of DNA

10. Removal of introns, noncoding regions of DNA

11. Involves the formation of a polypeptide chain

12. Occurs during interphase of the cell cycle

13. Requires DNA polymerase

14. Polyploidy results from

      (A) nondisjunction
      (B) aneuploidy
      (C) Down syndrome
      (D) infection with a virus
      (E) RNA processing

Questions 15–19

Choose from the terms below.

      (A) Gel electrophoresis
      (B) Restriction enzymes
      (C) Polymerase chain reaction
      (D) Recombinant DNA

15. Referred to as molecular scissors

16. Used to separate large molecules of DNA on the basis of their rate of movement through an agarose gel in an electric field

17. Taking DNA from two sources and combining them into one cell

18. Could cure juvenile diabetes

19. An automated technique by which a tiny piece of DNA can be rapidly amplified.

20. 97 percent of the human genome

    (A) does not code for anything and is known as junk
    (B) codes for about 30,000 genes
    (C) codes for about 3 billion genes
    (D) is located in the mitochondria
    (E) has been cut up by restriction enzymes

EXPLANATION OF ANSWERS

1. **(D)** Adenine bonds to thymine; cytosine bonds to guanine. The other choices are all correct statements about DNA.

2. **(B)** If guanine makes up 28 percent of the DNA, then there must be an equal amount of cytosine (28 percent), for a total of 56 percent. That leaves 44 percent for both adenine and thymine. Divide 44 by 2 to find the percentage of thymine in the DNA and the percentage of adenine in the sample of DNA, 22 percent.

3. **(D)** RNA contains the nitrogenous bases C, G, A, and U for uracil, in place of thymine. All the other statements about RNA are correct.

4. **(D)** Restriction enzymes cut DNA at specific restriction sites. They are necessary for genetic engineering. Introns are noncoding regions of DNA. Exons are expressed sequences or genes. PCR stands for polymerase chain reaction and is a lab technique for producing large amounts of DNA quickly.

5. **(E)** When DNA makes RNA, A (adenine) binds to U (uracil) and C (cytosine) binds with G (guanine). Uracil replaces thymine in RNA.

6. **(A)** RNA processing is a normal part of the synthesis of proteins by the cell. It always occurs after transcription but before the mRNA moves out of the nucleus to the ribosome to make proteins. It does not result in the formation of mutations. Since lengths of RNA (introns) are removed from the initial transcript, the final transcript is shorter than the initial transcript. All the other statements are correct about RNA processing.

7. **(B)** Transcription is the process by which DNA makes RNA.

8. **(B)** Transcription is the process by which DNA makes RNA. It occurs in the nucleus and requires RNA polymerase.

9. **(C)** Replication is the synthesis of a new molecule of DNA from one template DNA molecule. Replication occurs during interphase.

10. **(D)** Before a newly synthesized strand of RNA is shipped out of the nucleus to the ribosome, it is edited or processed, and introns are removed.

11. **(A)** Translation is the process by which the mRNA sequence is converted into an amino acid sequence. This occurs at the ribosome.

12. **(C)** Replication, the making of a new strand of DNA, occurs during the S phase (synthesis) of interphase of the cell cycle. The cell cycle consists of interphase (G_1, S, and G_2), mitosis, and cytokinesis.

13. **(C)** Replication, the making of a new strand of DNA, occurs during the S phase (synthesis) of the cell cycle. The cell cycle consists of interphase (G_1, S, and G_2), mitosis, and cytokinesis.

14. **(A)** Polyploidy means having extra sets of chromosomes, $3n$, $4n$, and so on, in a cell. It results from an error during meiosis I when homologous pairs of chromosomes fail to separate as they should. The process of chromosomes not separating correctly during meiosis is known as nondisjunction. Aneuploidy (any abnormal number of chromosomes) and the common form of Down syndrome (having an extra chromosome 21 in body cells) also result from nondisjunction. Infections do not cause polyploidy.

15. **(B)** Restriction enzymes are extracted from bacteria and are used as a basic biotechnology tool. They cut DNA at specific restriction sites, such as GAATTCC.

16. **(A)** Gel electrophoresis is commonly used to separate DNA. The sample of DNA must first be cut into pieces by restriction enzymes and then run through a gel.

17. **(D)** Recombinant DNA technology is also referred to as biotechnology or genetic engineering.

18. **(D)** People with juvenile diabetes have a nonfunctioning gene for the production of insulin. A cure could be achieved if scientists could replace the nonfunctioning gene with a proper gene. We do not yet have that capability.

19. **(C)** PCR can make millions of copies of DNA in only a few hours. It requires special heat-stable enzymes. Once the DNA has been amplified, the copies can be used in various ways in the biotechnology industry.

20. **(A)** The genome consists of all the genetic material in an organism. Most of the human genome does not code for anything. It is called junk and includes introns, and other noncoding regions of DNA. It is estimated that humans have 30,000 genes. The entire genome consists of about 3 billion base pairs. There is a small amount of DNA, only 16,000 base pairs, in mitochondria, but most DNA is located in the nucleus. Restriction enzymes are used as a tool for bioengineering; they cut DNA at special restriction sites. Restriction enzymes are extracted from bacteria and are not located inside human cells.

EVOLUTION AND DIVERSITY

Evolution

CHAPTER 11

- Basics of evolution
- Evidence of evolution
- Lamarck vs. Darwin and the theory of natural selection
- Types of natural selection
- Variation in a population
- Population stability—Hardy-Weinberg
- Isolation and the formation of new species
- Patterns of evolution
- Theories about evolution
- How life began
- Heterotroph hypothesis and theory of endosymbiosis
- Important concepts of evolution

An understanding of evolutionary theory helps scientists understand every field of biology from molecular biology to ecology. Evolution is the change in the genes of a population on Earth over time. **Microevolution** refers to the changes in one gene pool of a population over generations. **Macroevolution** refers to speciation, the formation of an entirely new species.

A critical thing to remember is that individuals never change or evolve. A population is the smallest group that can evolve. A population consists of all the members of one species in one place. For example, all the lions on the Masai plain in Kenya, or the Blue Jays in Madison, Wisconsin, are populations.

> **REMEMBER**
>
> Individuals never evolve; only populations evolve.

EVIDENCE OF EVOLUTION

Six areas of scientific study that provide evidence for evolution.

1. Fossil record
2. Comparative anatomy
3. Comparative biochemistry
4. Comparative embryology
5. Molecular biology
6. Biogeography

Fossil Record

The fossil record reveals the existence of species that have become extinct or have evolved into other species. The fossil record shows these important facts.

- 99 percent of all organisms that ever lived on Earth are now extinct.
- Through studies of radioactive dating and half-life, we know that Earth is about 4.6 billion years old.
- Prokaryotic cells are the oldest fossils and were the first organisms to develop on Earth.
- Paleontologists have discovered many **transitional fossils** that link older extinct fossils to modern species. For example, *Archaeopteryx* is a fossil that shows both reptile and bird characteristics. There also exist transition fossils that demonstrate that *Eohippus*, the ancient horse, is an ancestor of the modern horse, *Equus*.

Comparative Anatomy

Organisms that have similar anatomical structures are related to each other and share a common ancestor. For example, a comparison of dental structure in chimpanzees and humans demonstrates that we are related and that we both descended from a common ancestor less than 10 million years ago.

1. **Homologous structures.** Examples of homologous structures are the wing of a bat, the lateral fin of a whale, and the human arm. They all have the same internal bone structure, although the function of each varies. These structures have a common origin (all three are characteristics of mammals). Homologous structures are evidence of divergent evolution. (See "Patterns of Evolution" on page 161.)

2. **Analogous structures.** Analogous structures, such as a bat's wing and a fly's wing, have the same function but not the same underlying structure. The similarity is merely superficial and reflects adaptation to a similar environment. These structures are not evidence of a common origin or common ancestry. Analogous structures are evidence of convergent evolution. (See "Patterns of Evolution" on page 161.)

3. **Vestigial structures.** Vestigial structures, such as the appendix, are evidence that structurally, animals have evolved. The appendix is a vestige of a structure needed when our ancient ancestors ate a very different diet.

Comparative Biochemistry

Organisms that have a common ancestor will have common biochemical pathways. The more closely related organisms are to each other, the more similar their biochemistry is. Humans and mice are both mammals. This close relationship is the reason medical researchers can test new medicines on mice and extrapolate the results to humans.

Comparative Embryology

Closely related organisms go through similar stages in their embryonic development because they evolved from a common ancestor. For example, all vertebrate embryos go through a stage in which they have gill pouches on the sides of their throats. In fish, the gill pouches develop into gills. In humans, they develop into eustachian tubes that connect the middle ear with the throat.

Molecular Biology

Since all aerobic organisms contain cells that carry out respiration and require electron transport chains, they also all contain the necessary polypeptide, cytochrome *c*. A comparison of the amino acid sequence of cytochrome *c* among different organisms shows which are most closely related. The cytochrome *c* in human cells is identical to that of our closest relative, the chimpanzee, but differs somewhat from that of a pig and is vastly different from the cytochrome *c* found in paramecia or in oak leaves.

Biogeography

The theory of **continental drift** states that about 200 million years ago, the continents were locked together in a single supercontinent known as **Pangaea**, which slowly separated into seven continents over the course of the next 150 million years. Study of the location of the fossils, confirms the theory that marsupials migrated by land from South America across Antarctica to Australia before those two became separate continents about 55 million years ago. Today most of the world's marsupials are isolated in Australia.

LAMARCK VS. DARWIN

Lamarck was a contemporary of Darwin and also developed a theory of evolution. His theory relied on the ideas of inheritance of acquired characteristics and use and disuse. Lamarck stated that individual organisms change in response to their environment. The giraffe developed a long neck because it ate leaves of the tall acacia tree for nourishment and had to stretch to reach them. Lamarck believed that the animals stretched their necks and passed the acquired trait of an elongated neck onto their offspring. Although this theory may seem funny today, it was widely accepted in the early 19th century.

Darwin was a naturalist and author who, when he was 22, left England aboard the HMS *Beagle* to visit the Galapagos Islands, South America, Africa, and Australia. He collected, studied, and classified many organisms that no one in Europe had ever seen. His study of this amazing variety of organisms led him to develop his theory of natural selection, which explains how populations evolve and how new species develop. Darwin published "On the Origin of the Species" in 1859. It was a sensation. The first printing sold out on the first day it was published.

DARWIN'S THEORY OF NATURAL SELECTION

Here is the essence of Darwin's theory of Natural Selection.

- Populations tend to grow exponentially, to overpopulate, and to exceed their resources. Darwin developed this idea after reading **Malthus**, who published a treatise on population growth, disease, and famine in 1798.
- Overpopulation results in competition and a struggle for existence.
- In any population, there is variation and an unequal ability of individuals to survive and reproduce. Although everyone would agree that this is obviously so, at the time, no one understood genetics and therefore could not explain the source of the variation. Mendel's theory of genetics, which was published in 1865 (although not understood for 35 years), would have given Darwin a basis for understanding variation in a population. Another source of variation, and a major one, mutation, was not understood until Hugo de Vries described it in the early 1900s.
- Only the best-fit individuals survive and get to pass on their traits to offspring. This is commonly known as **survival of the fittest**. According to Darwin, the degree of fitness is measured by the ability of an individual to survive and to reproduce in its environment. Evolution occurs as advantageous traits accumulate in a population.

How the Giraffe Got Its Long Neck

According to Darwin's theory, ancestral giraffes were short-necked animals, although neck length varied from individual to individual. As the population of animals competing for the limited food supply increased, the taller individuals had a better chance of surviving than those with shorter necks. Over time, the proportion of giraffes in the population with longer necks increased until only long-necked giraffes existed. Remember that no individual animal's neck grew longer. The average length of the neck *in the population* changed.

How the Peppered Moth Changed from Light to Dark

Until 1845 in England, most peppered moths were light colored; few dark individuals could be found. With increasing industrialization, smoke and soot polluted the environment, making all the plants and rocks black. By the 1950s, all moths in the industrialized regions were dark; few light-colored individuals could be found. Before the industrial revolution, white moths were camouflaged in their environment while dark moths were easy prey for predators (birds). After the environment was darkened by heavy pollution, things shifted. Dark moths were camouflaged and had the selective advantage. Within a relatively short time, 100 years, dark moths replaced the light moths in the population. This darkening due to industrialization is referred to as industrial melanism. Remember, no single individual changed. Instead, the frequency of an allele (for color) in the population changed.

Evolution and Drug Resistance

Not all evolution occurs slowly. Natural selection can produce very rapid shifts in populations. For example, a few years after the discovery of antibiotics, bacteria appeared that were resistant to these drugs. The appearance of antibiotics did not induce mutations for resistance; it merely killed susceptible bacteria. Since only resistant individuals survive to reproduce, the next generation will all be resistant to the antibiotic they were exposed to. An entire population of bacterium can "become" resistant to a particular antibiotic in a matter of months. Once again, individual bacteria do not evolve; it is the population that evolves.

> **INTERESTING FACT**
>
> A new flu vaccine must be developed every year because the influenza virus evolves so rapidly.

The current treatment for AIDS (acquired immune deficiency disease) is a cocktail of drugs including AZT, which slows the progression of the disease. In some patients who have been taking the drugs for years, the virus that causes AIDS suddenly becomes resistant to these drugs and the patient quickly sickens. This is because the AIDS virus has the ability to mutate and evolve rapidly. The viruses that are susceptible to the drugs die, while those that have mutated and are resistant survive and reproduce an entire population that is drug resistant. For this reason, we have not been able to cure AIDS.

TYPES OF NATURAL SELECTION

The process of natural selection can alter the frequency of inherited traits in a population in three different ways, depending on which phenotypes in a population are favored. These types of selection are **stabilizing selection**, **diversifying** or **disruptive selection**, and **directional selection**.

Stabilizing Selection

Stabilizing selection eliminates the numbers of extremes and favors the more common intermediate forms. Many mutant forms are weeded out in this way. For example, stabilizing selection keeps the majority of birth weights in humans between 6 and 9 pounds. For babies much smaller or larger than this, mortality is greater. This is illustrated in Figure 11.1.

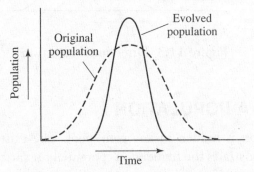

Figure 11.1 Stabilizing selection

Disruptive or Diversifying Selection

Disruptive selection increases the numbers of extreme types in a population at the expense of intermediate forms. A single population of snails can contain animals with either striped or plain shells. In the short term, this results in what is called *balanced polymorphism*, where two or more phenotypes coexist in a population. Over great lengths of time, disruptive selection may result in the formation of two entirely new species, as shown in Figure 11.2.

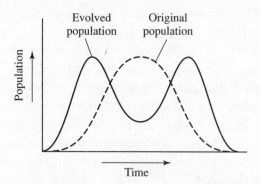

Figure 11.2 Disruptive selection

Directional Selection

Changing environmental conditions give rise to directional selection, where one phenotype replaces another in the gene pool. This was the case with peppered moths in England in the 20th century. Figure 11.3 diagrams this.

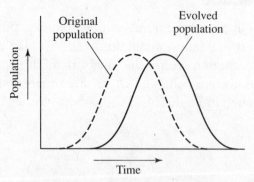

Figure 11.3 Directional selection

VARIATION IN A POPULATION

Tremendous variation is hidden in any gene pool. The existence of hundreds of breeds of dogs demonstrates the tremendous potential for variation within a species. Dogs such as the Great Dane, Chihuahua, and Beagle, which are all so very different from each other, belong to one species, *Canis familiaris*.

Although Darwin could not explain the origin of all the variation he saw in different populations, we now understand the process. The sources of variation in a population are **mutation**, **genetic drift**, and **gene flow**.

Mutation

Mutations are changes in genetic material and are the raw material for evolutionary change. A single **point mutation** can introduce a new allele into a population. These were first identified and named by the botanist Hugo de Vries in the early 1900s.

Genetic Drift

Genetic drift is change in the gene pool due to chance. There are two examples: the **bottleneck effect** and the **founder effect**.

BOTTLENECK EFFECT

Natural disasters such as fire, earthquake, and flood reduce the size of a population nonselectively, resulting in a loss of genetic variation. The resulting population is much smaller and not representative of the original one. Certain alleles may be under or overrepresented compared with the original population. This phenomenon is known as the bottleneck effect.

THE FOUNDER EFFECT

When a small population breaks away from a larger one to colonize a new area, it is most likely not genetically representative of the original larger population. Rare alleles may be overrepresented. This is known as the founder effect. It occurred in the Old Order of Amish of Lancaster, Pennsylvania, all of whom descended from a small group of settlers who came to the United States from Germany in the 1770s. Apparently, one or more of the settlers carried the rare but dominant gene for polydactyly, having extra finger and toes. Due to the extreme isolation and intermarriage of the close community, this population now has a high incidence of polydactyly.

Gene Flow

Gene flow is the movement of alleles into or out of a population. It can occur as a result of the migration of fertile individuals or gametes between populations. For example, pollen from one valley can be carried by the wind across a mountain to another valley.

POPULATION STABILITY— HARDY-WEINBERG EQUILIBRIUM

Hardy and Weinberg, two scientists, developed a theorem that described a stable, nonevolving population, that is, one in which allelic frequency does not change. For example, if the frequency of an allele for a particular trait is 0.5 and the population is not evolving, then in 1,000 years, the frequency of that allele will still be 0.5. This is called Hardy-Weinberg equilibrium. According to Hardy-Weinberg, if the population is stable, the following must be true.

1. **The population must be very large.** In a small population, the smallest change in the gene pool will have a major effect in allelic frequencies. In a large population, a small change in the gene pool will be diluted by the sheer number of individuals and there will be no change in the frequency of alleles.

> **STUDY TIP**
>
> Make sure you know the characteristics of a stable population according to Hardy-Weinberg.

2. **The population must be isolated from other populations.** There must be no migration of organisms into or out of the gene pool because that could alter allelic frequencies.

3. **There must be no mutations in the population.** A mutation in the gene pool could cause a change in allelic frequency or introduce a new allele.

4. **Mating must be random.** If individuals select mates, then those individuals who are better fit will have a reproductive advantage and the population will evolve.

5. **There must be no natural selection.** Natural selection causes changes in relative frequencies of alleles in a gene pool.

Hardy-Weinberg Equation

The Hardy-Weinberg equation enables us to calculate frequencies of alleles in a population. Although it can be applied to complex situations of inheritance, for the purpose of explanation here, we will discuss a simple case—a gene locus with only two alleles. Scientists use the letter p to stand for the dominant allele and the letter q to stand for the recessive allele.

The Hardy-Weinberg equation is

$$p + q = 1 \quad \text{or} \quad p^2 + 2pq + q^2 = 1$$

Look at the monohybrid cross as a basis for this equation.

	A	a
A	AA	Aa
a	Aa	aa

$$p^2 = AA; \ 2pq = 2(Aa); \ q^2 = aa$$

STUDY TIP

p = dominant allele

q = recessive allele

STUDY TIP

p^2 represents the homozygous dominant individual.

q^2 represents the homozygous recessive individual.

SAMPLE PROBLEM

Take a pencil and do the problem along with the text.

If 9 percent of the population has blue eyes, what percent of the population is hybrid for brown eyes? Homozygous for brown eyes?

To solve this problem, follow these steps:

1. The trait for blue eyes is homozygous recessive, *bb*, and is represented by q^2. In this example, $q^2 = 9\%$

   Convert to a decimal: $q^2 = 0.09$

2. To solve for q, take the square root of $0.09 = 0.3$

3. Since $p + q = 1$ and $q = 0.3$, then $p = 0.7$

4. The hybrid condition is represented by $2pq$

5. To solve for the percent of the population that is hybrid, substitute for $2(p)(q)$.

    $2(0.7)(0.3) = 0.42$.

    Convert to a percent: $2pq = 42\%$

6. Homozygous dominant is represented by p^2.

    $p^2 = (0.7)^2 = 0.49$

    Convert to a percent: $p^2 = 49\%$

ISOLATION AND NEW SPECIES FORMATION

A **species** is a population whose members have the potential to interbreed in nature and produce viable, fertile offspring. For example, lions and tigers do not belong to the same species because, although they can be induced to interbreed in captivity, they would not naturally do so. Horses and donkeys do interbreed in nature. However, the offspring that result is a mule, which is not fertile. Therefore, the horse and donkey belong to different species.

Anything that fragments a population and isolates small groups of individuals may foster the formation of new species. This is true because isolated populations are subject to different selective pressures in their respective environments. If enough time elapses, the two populations may become so different that, even if they were brought back together, interbreeding would not occur. At this point, a new species is said to have come into being. Figure 11.4 shows two examples of isolating factors.

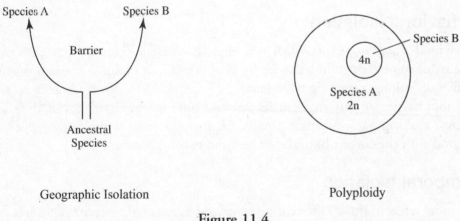

Geographic Isolation   Polyploidy

Figure 11.4

Six different forms of isolation commonly cause a new species to form:

1. Geographic isolation
2. Polyploidy
3. Habitat isolation
4. Behavioral isolation
5. Temporal isolation
6. Reproductive isolation

Geographic Isolation

Geographic isolation occurs when species are separated. Mountain ranges, canyons, rivers, lakes, or glaciers may cause significant isolation between species.

Polyploidy

Polyploidy is a type of mutation that results from errors during meiosis. Instead of being monoploid (n) or diploid ($2n$), polyploid organisms can be tetraploid ($4n$) or octoploid ($8n$). Nearly half of all flowering plants and the vast majority of ferns are polyploid. Polyploid organisms cannot breed with organisms that are not polyploid and therefore are isolated from them. This is seen in Figure 11.4.

Habitat Isolation

Habitat isolation occurs when two organisms live in the same area but encounter each other rarely. Two species of one genus of snake can be found in the same geographic area, but one inhabits the water while the other is mainly terrestrial.

Behavioral Isolation

Behavioral isolation occurs when two animals become isolated from each other because of some change in behavior by one member or group. For example, male fireflies of various species signal to females of their kind by blinking lights on their tails in a particular pattern. Females respond only to characteristics of their own species, flashing back to attract males. If, for any reason, the female does not respond with the correct blinking pattern, no mating occurs.

Temporal Isolation

Temporal refers to time. Different plants of one species living in the same area may become functionally separated into two populations via temporal isolation because some plants become sexually mature earlier and begin to flower in the cooler part of the season while other plants flower in the later, warmer part of the growing season.

Reproductive Isolation

Closely related species may be unable to mate because of anatomical incompatibility. For example, a small male dog and a large female dog cannot mate because of the enormous size differences between the two animals.

PATTERNS OF EVOLUTION

How species evolve is classified into five patterns: divergent, convergent, parallel, coevolution, and adaptation radiation. These are illustrated in Figure 11.5.

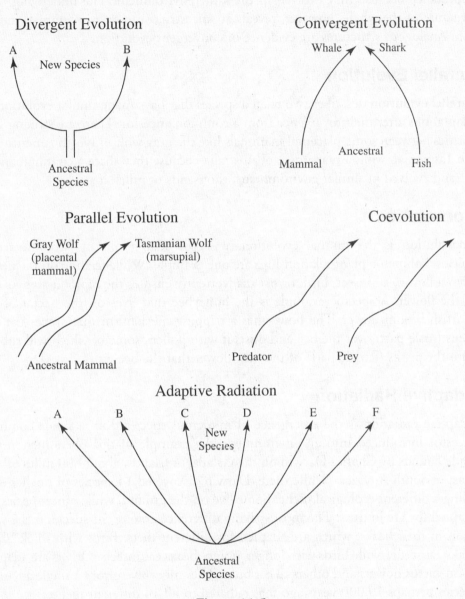

Figure 11.5

Divergent Evolution

Divergent evolution occurs when a population becomes isolated (for any reason) from the rest of the species and becomes exposed to new selective pressures, causing it to evolve into a new species. Homologous structures are evidence of divergent evolution.

Convergent Evolution

When unrelated species occupy the same environment, they are subjected to similar selective pressures and show similar adaptations. The classic example of **convergent evolution** is the whale (a mammal) and the fish. Both have a streamlined appearance because they evolved in the same environment. The underlying bone structure of the whale, however, reveals an ancestry common to mammals, not to fish. Analogous structures are evidence of convergent evolution.

Parallel Evolution

Parallel evolution describes two related species that have made similar evolutionary adaptations after their divergence from a common ancestor. There are striking similarities between some placental mammals like the gray wolf of North America and the Tasmanian wolf, a marsupial, of Australia because they share a common ancestor and evolved in similar environments, thousands of miles apart.

Coevolution

Coevolution is the mutual evolutionary set of adaptations of two interacting species. Pollinator-plant relationships are one example. While feeding on the nectar from a flower, an insect, bird, or bat inadvertently ensures the reproductive success of the flower. A specific example is the honeybee that lives on the nectar of the Scottish broom flower. The flower has a tripping mechanism that arches the stamens (male part) over the bee and dusts it with pollen, some of which will rub off onto the pistils (female part) of the next flower that the bee visits.

Adaptive Radiation

Adaptive radiation is the emergence of numerous species from a single common ancestor introduced into an environment. An example of this phenomenon was made famous by Charles Darwin on the Galapagos Islands, about 600 miles off the coast of South America. While there, Darwin discovered 14 species of finches each filling a different ecological niche. Some live on the ground, while other species are adapted for life in trees. The most striking difference among the species is the variation in their beaks, which are adapted for different diets. Birds with thick, short beaks eat seeds, while birds with longer, pointy beaks eat insects. Others are adapted to eat cactus flowers and others to eat buds. They all evolved from a single ancestral species perhaps 10,000 years ago that radiated to fill 14 different niches.

THEORIES ABOUT EVOLUTION

Several theories about evolution have been proposed. Some have been proven to be wrong.

Gradualism

Gradualism is the theory that organisms descend from a common ancestor gradually, over a long period of time, in a linear or branching fashion. Big changes occur by an accumulation of many small ones. According to this theory, fossils should exist as evidence of every stage in the evolution of every species with no missing links.

However, the fossil record is at odds with this theory because scientists rarely find transitional forms or missing links. Scientists have abandoned this theory for the theory of punctuated equilibrium.

Punctuated Equilibrium

The favored theory today is called **punctuated equilibrium**. It was developed by Stephen J. Gould and Niles Eldridge after they observed that Darwin's theory of gradualism was not supported by the fossil record. This theory proposes that new species appear suddenly after long periods of stasis. Most likely, a new species arises in a different place and expands its range, competing with and replacing the ancestral species that becomes extinct. Figure 11.6 illustrates both gradualism and punctuated equilibrium.

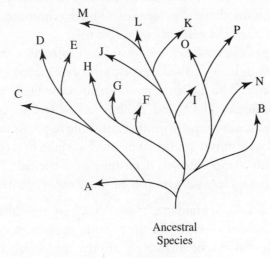

Gradualism

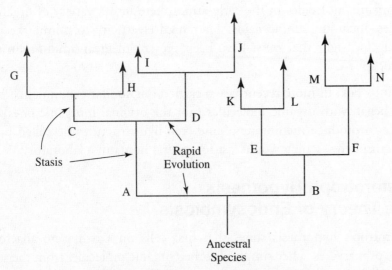

Figure 11.6 Punctuated equilibrium

Spontaneous Generation

Spontaneous generation is the theory that living things emerge from nonliving or inanimate objects. This belief was disproved by two scientists. In the 17th century, Francesco Redi performed a now-famous experiment in which he put decaying meat into a group of wide-mouthed jars—some covered by lids, some covered by fine cheesecloth, and some left open. He demonstrated that maggots arose only where flies were able to lay their eggs.

In the early 1860s, Louis Pasteur, a microbiologist used a goose-necked flask of his own making to prove that microorganisms appeared only as contaminants from the air and not spontaneously.

HOW LIFE BEGAN

All organisms alive today represent less than 3 percent of all the organisms that ever lived. This means that more than 97 percent of all life has become extinct. The story of evolution based on scientific evidence goes like this.

Radioisotope dating of our solar system, including Earth, puts the age at about 4.6 billion years. After the big bang formed the planets, Earth's outer surface cooled and solidified to form a crust. The ancient environment most likely consisted of CH_4 (methane), NH_3 (ammonia), H_2O, and N_2 but lacked free oxygen. Intense heat, lightning, and U.V. radiation in the primitive atmosphere provided the energy for a multitude of chemical reactions that ultimately produced the first cell. Scientists have tried to mimic this early atmosphere to determine how the first organic molecules and earliest life developed. Here is a synopsis of those experiments.

> **REMEMBER**
>
> There was no free oxygen in the ancient Earth's atmosphere.

1. A. I. Oparin and J. B. S. Haldane, in the 1920s, hypothesized separately that under the conditions of early Earth, organic molecules could form. They stated that in the absence of corrosively reactive molecular oxygen that would react with and degrade them, organic molecules could form and persist.

2. Stanley Miller and Harold Urey, in the 1950s, tested the Oparin-Haldane hypothesis and proved that almost any energy source would have converted inorganic molecules in the early atmosphere into a variety of organic molecules, including amino acids. They used electricity to mimic the lightning and U.V. light that must have been present in great amounts in the early atmosphere.

3. Sidney Fox, in more recent years, carried out similar experiments. However, he began with organic molecules (not the original inorganic ones) and was able to produce membrane-bound, cell-like structures he called proteinoid microspheres, which would last for several hours in a laboratory.

The Heterotroph Hypothesis and the Theory of Endosymbiosis

The **heterotroph hypothesis** states that first cells on Earth were anaerobic heterotrophic prokaryotes. They simply absorbed organic molecules from the surrounding primordial soup to use as a nutrient source. The fossil record demonstrates that the first cell evolved about 3.5 billion years ago.

Eukaryotic cells with a nucleus and other internal organelles evolved about 1.5 billion years ago from prokaryotic cells. This occurred as tiny bacteria took up residence inside larger prokaryotic cells and performed important functions for the host cell. These mutually beneficial symbiotic relationships resulted in nuclei, chloroplasts, and mitochondria. This theory of endosymbiosis was developed by Dr. Lynn Margulis.

According to the fossil record, the first multicellular animals appeared about 565 million years ago. Then over a span of just 40 million years (a blink of an eye in geologic terms), virtually every major phylum of animal appeared. This period of time in which so many animals appeared is known as the Cambrian explosion. Animals moved from the oceans to the land, filling every available niche as competition for limited resources increased in the oceans and as they evolved the traits necessary to live in a dry environment. Modifications for life on land evolved in both animals and plants.

Several characteristics enabled animals to move to land:

- Lungs
- Skin to keep the animals from drying out
- Limbs to move about
- Mechanisms for internal fertilization or
- Shell to protect their eggs and to keep them from drying out

Several characteristics enabled plants to move to land:

- Roots that anchor them into the soil and absorb water
- Supporting cells to enable them to compete favorably for light
- Vascular tissue to carry water upward
- Waxy molecule (cutin) to protect the leaves from dehydrating
- Seeds, a protective package for the embryo and its food

Mammals appeared about 210 million years ago. Primates, including apes, appeared about 25 million years ago. Humans did not evolve from apes; we both evolved from a common ancestor about 5 million years ago. Human ancestors arose in Africa. How they spread throughout the rest of the world is not agreed upon. However, scientists agree that modern humans, *Homo sapiens*, arose about 100,000 years ago.

IMPORTANT CONCEPTS OF EVOLUTION

1. **Evolution is not always a slow process.** A species of bacteria can develop resistance to a particular antibiotic after only a few months of exposure.

2. **Evolution does not occur at the same rate in all organisms.** Humans have changed a great deal in the last 100,000 years, whereas the horseshoe crab has hardly changed at all.

3. **Evolution does not always cause organisms to become more complex.** Instead, it may act in such a way that complex forms give rise to simpler ones. The early embryo of the sea star is bilaterally symmetrical, which is similar to human development and considered to be advanced and complex. The adult sea star, however, shows radial symmetry, which is considered simple and primitive.

4. **Evolution occurs in populations, not individuals.** A single giraffe did not develop a long neck because it needed it. Instead, short-necked giraffes could not compete in a competitive environment and died out. Only long-necked giraffes survived.

5. **Evolution is directed by changes in the environment.** Animals that evolved in the ocean must be streamlined in order to move freely.

MULTIPLE-CHOICE QUESTIONS

1. All of the following statements about evolution are true EXCEPT

   (A) evolution occurs more rapidly at some times than at others
   (B) evolution is directed by changes in the environment
   (C) evolution is not always from the simple to the complex but may act in such a way that complex forms give rise to simpler ones
   (D) single organisms, as well as populations, do commonly evolve
   (E) evolution is not always a slow, gradual process

2. According to the Hardy-Weinberg theory, which of the following represents a homozygous dominant individual?

   (A) p
   (B) q
   (C) p^2
   (D) q^2
   (E) $2pq$

3. All of the following are part of the Hardy-Weinberg theorem, which describes a stable, nonevolving population, EXCEPT

   (A) natural selection exists
   (B) mating must be random
   (C) no mutations can occur in the population
   (D) the population must be isolated from others
   (E) the population must be large

4. Which of the following is a transition fossil?

   (A) Reptiles
   (B) Amphibians
   (C) *Archaeopteryx*
   (D) Fish
   (E) Apes

5. All of the following are part of Darwin's theory of natural selection
   EXCEPT

   (A) populations tend to overpopulate
   (B) in any population, there is unequal ability among individuals to
        survive and to reproduce
   (C) only the best-fit individuals survive and get to pass on their genes
   (D) overpopulation leads to a struggle for existence
   (E) organisms change in response to a need in the environment

6. All the following are true of homologous structures EXCEPT

   (A) they demonstrate common ancestry
   (B) an example is the wing of the bat and a person's arm
   (C) an example is the wing of the bat and the wing of a fly
   (D) they are not the same thing as analogous structures
   (E) an example is the whale's lateral fin and a person's arm

Questions 7–12

Choose from the list of terms below.

   (A) Charles Darwin
   (B) Lamarck
   (C) Stephen J. Gould and Niles Eldridge
   (D) Oparin and Haldane
   (E) Hardy-Weinberg

7. Inheritance of acquired traits

8. Use and disuse

9. Theory of punctuated equilibrium

10. Populations tend to overpopulate

11. Hypothesized that under conditions that existed in early Earth, organic
    molecules could form

12. Survival of the fittest

13. The age of Earth according to scientific evidence is closest to

   (A) 4,000 years old
   (B) 600 years old
   (C) 6,000 years old
   (D) 4 million years old
   (E) 4 billion years old

14. Pangaea is a

   (A) transition fossil
   (B) molecule that is commonly analyzed and used to show which organisms are related
   (C) species of bacteria that is resistant to all antibiotics
   (D) vestigial structure
   (E) single supercontinent that existed 200 million years ago

15. Which of the following is an example of divergent evolution?

   (A) Wildebeests separated from each other by a newly formed river are now separate species.
   (B) Whales and fish have a streamlined appearance because they evolved in the same environment.
   (C) Insects and the flowers they pollinate have evolved together over millions of years.
   (D) Polydactyly, having extra fingers, is common in the Amish of Pennsylvania.
   (E) Change occurs in the gene pool due to chance.

Questions 16–20

Choose from the terms below.

   (A) Geographic isolation
   (B) Polyploidy
   (C) Reproductive isolation
   (D) Adaptive radiation
   (E) Directional selection

16. Darwin discovered 14 species of finches on the Galapagos Islands that all evolved from one original species of finch

17. Having extra sets of chromosomes in every cell

18. The peppered moths are one example

19. Two populations of one species evolved into two separate species after being separated for millions of years by a canyon

20. A very small dog and a very large dog cannot mate because of the enormous size difference of the two animals

EXPLANATION OF ANSWERS

1. (D) The population is commonly the smallest group to evolve. In the case of the peppered moths, for example, no single white moth became dark. When the environment became darker, the white moths were visible and were preyed upon, while the dark ones had the selective advantage.

2. (C) The dominant allele is represented by p, and p^2 is the homozygous dominant individual. The recessive allele is represented by q, and q^2 is the homozygous recessive individual. $2pq$ represents the hybrid individual.

3. (A) If natural selection were operating, then the population would be evolving. Hardy-Weinberg describes a population in equilibrium, in other words, not evolving. All the other choices describe populations that are stable and not evolving.

4. (C) *Archaeopteryx* is a transition fossil that is part reptile and part bird. Another example of a transition fossil is *Eohippus*. It is an ancient horse, a transition between earlier horses and *Equus*, the modern species of horse. The other choices are names of living animals.

5. (E) Change in an organism can be caused by mutations. However, mutations are random and do not result from a need to change. Organisms that are no longer adapted to a changing environment either die or migrate. Choice (E) best describes statements made by Lamarck, who stated that organisms change because they need to. All the other statements are correct about the theory of natural selection.

6. (C) Although the wing of the bat and the fly's wing have the same function, they are not alike structurally. Therefore, they are analogous structures, not homologous. They do not demonstrate a common ancestry.

7. (B) Lamarck's theory of evolution relied on the ideas of use and disuse and of inheritance of acquired characteristics. He was a contemporary of Darwin, but his theories were disproved long ago.

8. (B) Lamarck's theory of evolution relied on the ideas of use and disuse and of inheritance of acquired characteristics. He was a contemporary of Darwin, but his theories were disproved long ago.

9. (C) The theory of punctuated equilibrium proposes that new species appear suddenly after long periods with no evolution. It replaced the long-held theory of gradualism described by Darwin.

10. (A) Populations tend to grow exponentially, to overpopulate, and to exceed their resources. Darwin developed this idea after reading Malthus, who published a treatise on population growth, disease, and famine in 1798.

11. **(D)** Oparin and Haldane, in the 1920s, hypothesized separately that under the conditions of early Earth, organic molecules could form. They stated that in the absence of corrosively reactive molecular oxygen that would react with and degrade them, organic molecules could form and persist.

12. **(A)** This is one of four basic tenets of Darwin's theory of evolution.

13. **(E)** Through studies of radioactive dating and half-life, we know that Earth is about 4.6 billion years old.

14. **(E)** The theory of continental drift states that about 200 million years ago the continents were locked together in a single supercontinent known as Pangaea. It slowly separated into seven continents over the course of the next 150 million years. The molecule described in choice (B) is cytochrome *c*. Many species of bacteria have developed resistance to specific antibiotics, but few are resistant to all of them. An example of a vestigial structure is the appendix. An example of a transition fossil is *Archaeopteryx*.

15. **(A)** Choice (B) refers to convergent evolution, where unrelated organisms have a similar appearance because they evolved in the same or similar environments. Choice (C) refers to coevolution. Choice (D) refers to the founder effect, an example of genetic drift (a change in the gene pool due to chance). Choice (E) refers to genetic drift (a change in the gene pool due to chance).

16. **(D)** Darwin's finches are a classic example of adaptive radiation.

17. **(B)** Polyploidy, $2n$, $3n$, or more, is a type of mutation that results from errors during meiosis. It occurs commonly in flowers. Organisms that exhibit polyploidy cannot breed with organisms that are diploid ($2n$).

18. **(E)** What happened to the peppered moths in England is a good example of directional selection because the population changed from white to dark as time passed.

19. **(A)** Anything that fragments a population and isolates small groups of individuals may cause the formation of new species.

20. **(C)** Anything that prevents organisms of the same species from mating can cause the formation of new species because the two groups cannot share genes.

Taxonomy

CHAPTER 12

- Three-domain system of classification
- Kingdoms: Protista, Fungi, Plantae, and Animalia
- Evolutionary trends in animals
- Characteristics of animals
- Characteristics of mammals and primates

Taxonomy is a system by which we name and classify all organisms, living and extinct. The system we use today is based on the system developed in the 18th century by Carl Linnaeus (Carl von Linné). It is known as the system of **binomial nomenclature** because every organism has a two-part name. For example, human is *Homo sapiens* and lion is *Panthera leo*. In addition, Linnaeus classified every organism into a hierarchy of **taxa**, or levels of organization. These taxa are kingdom, phylum, class, order, family, genus, and species. Kingdom is the most general, consisting of the most varied organisms, and species is the most specific, consisting of organisms that are the most similar.

In the 20th century, our system of classification went through many changes. Prior to the 1950s and 1960s, all organisms were placed into only three kingdoms. From the 1960s to around 1990, scientists classified all organisms into five kingdoms: Monera, Protista, Fungi, Plantae, and Animalia. In 1990, some scientists added a sixth kingdom, the Archaebacteria. This included **extremophiles**, microorganisms that seemed so different from bacteria that they had to be placed into a separate kingdom.

Today, however, most scientists use another system, based on DNA analysis, which more accurately reflects evolutionary history and the relationships among organisms. This system is called the three-domain system. All life is organized into three **domains**, Bacteria, Archaea, and Eukarya, superkingdoms that include four of the original kingdoms. (The kingdom **Monera** is no longer used because in this system, prokaryotes are spread across two different domains, Archaea and Bacteria.)

The change to the three-domain system was necessary for one major reason: Archaea have so little in common with bacteria that they must have their own group. In addition, the name Archaebacteria had to be changed to Archaea because the Archaea are not bacteria, as seen in Figure 12.1. This sounds like a big change, but it is really simple.

> **REMEMBER**
>
> The term "Monera" is no longer used.

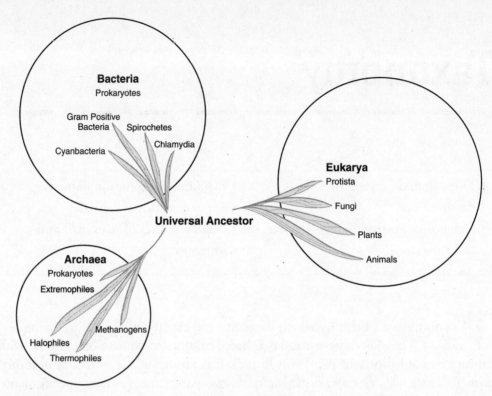

Figure 12.1

THE THREE-DOMAIN CLASSIFICATION SYSTEM

All organisms are classified into one of the following three domains.

Domain Bacteria

- All are single-celled **prokaryotes** with no internal membranes (no nucleus, mitochondria, or chloroplasts).
- Some are anaerobes; some are aerobes.
- Bacteria play a vital role in the ecosystem as **decomposers** that recycle dead organic matter.
- Many are **pathogens**, disease causing.
- Bacteria play a vital role in **genetic engineering**. The bacteria from the human intestine, *Escherichia coli*, are used to manufacture human insulin.
- Some bacteria carry out **conjugation**, a primitive form of sexual reproduction where individuals exchange genetic material.
- Bacteria have a thick, rigid cell wall containing a substance known as peptidoglycan.
- Some carry out photosynthesis, but others do not.
- There are no introns (noncoding regions within the DNA).
- Member species correspond roughly to the old grouping Eubacteria and include blue-green algae, bacteria like *E. coli* that live in the human intestine, those that cause disease like *Clostridium botulinum* and *Streptococcus*, and those necessary in the nitrogen cycle, like nitrogen-fixing bacteria, among others.
- Viruses are placed here because we do not know where else to place them.

Domain Archaea

- Unicellular
- Prokaryotic—no internal membranes such as a nucleus
- Includes extremophiles, organisms that live in extreme environments, like

  1. **Methanogens.** Obtain energy in a unique way by producing methane from hydrogen
  2. **Halophiles.** Thrive in environments with high salt concentrations like Utah's Great Salt Lake
  3. **Thermophiles.** Thrive in very high temperatures, like in the hot springs in Yellowstone Park or in deep-sea hydrothermal vents

- Introns present in some genes

Domain Eukarya

- All organisms have a nucleus and internal organelles.
- Eukarya includes the four remaining kingdoms: Protista, Fungi, Plantae, and Animalia.

DON'T FORGET

All animals belong in the domain Eukarya.

THE FOUR KINGDOMS OF EUKARYA

For the SAT Subject Test in Biology, you should have a general knowledge of the characteristics of organisms in each domain, each of the four kingdoms, and the nine common animal phyla as well as sample organisms belonging to each. For more details about plants, see the separate chapter "Plants."

Although the kingdom Monera may still appear in some older textbooks, it is no longer in use. The term is obsolete. As previously described, the prokaryotes are now classified in two different domains, Bacteria and Archaea.

Kingdom Protista

- This kingdom includes the widest variety of organisms, but all are **eukaryotes**.
- Most are single-celled, but many are primitive multicelled organisms.
- It includes **heterotrophs** and **autotrophs**.
- Examples of heterotrophs are amoeba and paramecium.
- Examples of autotrophs are euglenas, which have a red eyespot to locate light and chlorophyll to carry out photosynthesis.
- Protista move by various means: amoeba uses pseudopods; paramecium uses cilia; euglena uses a flagellum.
- Protista includes organisms that do not fit into the fungi or plant kingdoms, such as seaweeds and slime molds.
- Some protista sometimes carry out conjugation, a primitive form of sexual reproduction where individuals exchange genetic material, such as paramecium and algae.
- Some cause serious diseases, like amoebic dysentery and malaria.

Kingdom Fungi

- All are **heterotrophic eukaryotes**.
- They can be either unicellular or multicellular.
- Fungi carry out extracellular digestion by secreting hydrolytic enzymes outside the body. After digestion, the building blocks of the nutrients are absorbed into the body of the fungus by diffusion.
- They are important in the ecosystem as decomposers.
- Fungi are **saprobes**, organisms that obtain food from decaying organic matter. As such, they recycle nutrients in an ecosystem.
- Their cell walls are composed of **chitin**, not cellulose.
- Certain fungi combine with algae in a mutualistic, symbiotic relationship forming various lichens, which are photosynthetic. Lichens can survive harsh, cold environments and even live on bare rock. Lichens are often the **pioneer organisms**, the first to colonize a barren environment in an ecological succession.
- They reproduce asexually by budding (yeast), spore formation (bread mold), or fragmentation whereby a single parent breaks into parts that regenerate into whole new individuals.
- They also reproduce sexually.
- Examples include yeast, mold, mushrooms, and the fungus that causes athlete's foot.

Kingdom Plantae

- All are multicellular, nonmotile, autotrophic eukaryotes.
- Their cell walls are made of cellulose.
- Plants carry out photosynthesis using chlorophyll *a* and *b*.
- Plants store their carbohydrates as starch.
- They reproduce sexually by alternating between **gametophyte** (*n*) and **sporophyte** (2*n*) generations (known as alternation of generations).
- Some plants have vascular tissue (tracheophytes), and some have no vascular tissue (bryophytes).
- Examples include mosses, ferns, and cone-bearing and flowering plants.
- See the separate chapter "Plants" for more details.

Kingdom Animalia

- All are heterotrophic, multicellular eukaryotes.
- Most are motile, can move on their own.
- Most animals reproduce sexually with a dominant diploid (2*n*) stage.
- In most species, a small flagellated sperm fertilizes a larger, nonmotile egg.
- The traditional way of classifying animals is primarily based on anatomical features (homologous structures) and embryonic development.
- They are grouped in 35 phyla, but we commonly discuss 9: porifera, cnidarians, platyhelminthes, nematodes, annelids, mollusks, arthropods, echinoderms, and chordates.

EVOLUTIONARY TRENDS IN ANIMALS

Organisms began as tiny, primitive, single-celled organisms that lived in the oceans. The first multicellular eukaryotes evolved about 1.5 billion years ago. The appearance of each phylum of animal represents the evolution of a new and successful body plan. These important trends include specialization of tissues, germ layers, body symmetry, cephalization, and body cavity formation.

Specialized Cells, Tissues, and Organs

Begin with some definitions.

1. The **cell** is the basic unit of all forms of life. A neuron is a cell.

2. A **tissue** is a group of similar cells that perform a particular function. The sciatic nerve is a tissue.

3. An **organ** is a group of tissues that work together to perform related functions. The brain is a tissue.

Sponges (porifera) consist of a loose federation of cells, which are not considered tissue because the cells are relatively unspecialized. They possess cells that can sense and react to the environment but have no real nerve or muscular tissue.

Cnidarians like the hydra and jellyfish possess only the most primitive and simplest forms of tissue.

As larger and more complex animals evolved, specialized cells joined to form real tissues, organs, and organ systems. Flatworms have organs but no organ systems.

More complex animals, like annelids (earthworms) and arthropods (grasshoppers) have organ systems.

Germ Layers

Germ layers are the main layers that form various tissues and organs of the body. They are formed early in embryonic development and include the ectoderm, endoderm, and mesoderm.

1. The **ectoderm**, or outermost layer, becomes the skin and nervous system, including the nerve cord and brain.

2. The **endoderm**, the innermost layer, becomes the viscera (guts) or the digestive system.

3. The **mesoderm**, middle layer, becomes the blood, muscles, and bones.

Animals with only two cell layers, the porifera and cnidarians, are called diploblastic. Their bodies consist of ectoderm, endoderm, and **mesoglea** (middle glue), which holds the two layers together. The more complex animal phyla are **triploblastic**, having three true cell layers.

> **STUDY TIP**
>
> You must know which layer develops into which structures.

Bilateral Symmetry

Whereas primitive animals exhibit radial symmetry, sophisticated animals exhibit bilateral symmetry. Echinoderms are an exception because they are an advanced phylum and exhibit bilateral symmetry, as larvae but revert to radial symmetry as adults. In bilateral symmetry, the body is organized along a longitudinal axis with right and left sides that mirror each other. Most bilaterally symmetrical animals are triploblastic, with ectoderm, mesoderm, and endoderm.

Cephalization

Along with bilateral symmetry comes the development of a head end—the anterior, and a rear end—the posterior. Sensory apparatus and a brain, or simply ganglia, are clustered at the anterior end. Digestive, excretory, and reproductive structures are located at the posterior end. This enables animals to move faster to flee or to capture prey successfully. Simple animals—sponges and cnidarians—do not have a head end. More sophisticated animals, beginning with flatworms and ending with chordates, all show cephalization.

Coelom

The coelom is a fluid-filled body cavity that is completely surrounded by mesoderm tissue as seen in Figure 12.2. It represents a significant advance in the course of animal evolution because it provides a space for elaborate organ systems, like the digestive tract or cardiovascular system. Major organs could not have evolved without a coelom between the germ layers. Primitive animals, the flatworms, do not have a coelom and are known as **acoelomates**. Nematodes or roundworms are called **pseudocoelomates**; they have a fluid-filled tube between the endoderm and the mesoderm. (It is not completely lined by mesoderm.) **Coelomates** are animals with a coelom and are the most complex in the kingdom. These include the following phyla: Annelida, Mollusca, Arthropoda, and Chordata.

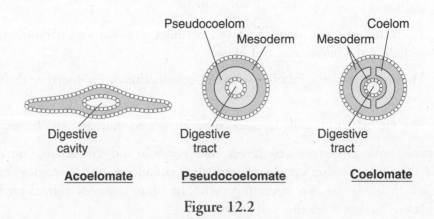

Figure 12.2

Figure 12.3 shows a family tree of animals organized by phylum and evolutionary trends. Table 12.1 identifies trends in animal development.

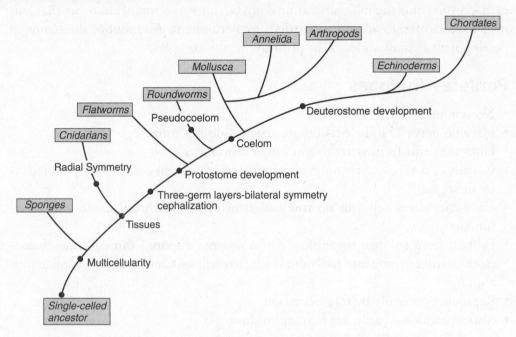

Figure 12.3 Trends in animal development

TABLE 12.1

Trends in Animal Development	
From the Primitive	**To the Complex**
No symmetry or radial symmetry	Bilateral symmetry
No cephalization	Cephalization with sensory apparatus
Mesoglea holds two cells layers together	Three cells layers, including mesoderm
Acoelomate	Pseudocoelomate or coelomate
No true tissues	Tissues, organs, and organ systems
Little specialization	Much specialization
Sessile	Motile

NINE COMMON ANIMAL PHYLA

You should be familiar with the nine common phyla and representative animals of each. As you study the individual animal phyla, think in terms of strategies that animals have evolved to adapt to a particular environment. Also, notice the trends in development in animals from the primitive to the complex.

Porifera—Sponges

- No symmetry
- Have no nerve or muscle tissue, are **sessile**—do not move
- Filter nutrients from water drawn into a central cavity
- Consist of two cell layers only: ectoderm and endoderm connected by noncellular mesoglea
- Have specialized cells but no true tissues or organs, each cell carries out many functions.
- Evolved from colonial organisms; if you squeeze a sponge through fine cheesecloth, it will separate into individual cells that will spontaneously reaggregate into a sponge
- Reproduce asexually by **fragmentation**
- Also reproduce sexually, are **hermaphrodites**

Cnidarians—Hydra and Jellyfish

- Radial symmetry
- Body plan is the **polyp** (vase shaped) or **medusa** (upside-down bowl shaped)
- Life cycle—some go through a planula larva (free-swimming) stage then go through two reproductive stages: asexually reproducing (polyp) and sexually reproducing (medusa)
- Two cell layers: ectoderm and endoderm connected by noncellular mesoglea
- Have a **gastrovascular cavity** where extracellular digestion occurs
- Carry out intracellular digestion inside body cells in **lysosomes**
- Have no transport system because every cell is in direct contact with the environment
- All members have stinging cells—**cnidocytes**—containing stingers, which are called **nematocysts**

Platyhelminthes—Flatworms Including Tapeworms

- They are the simplest animals with bilateral symmetry, an anterior end and three distinct cell layers: ectoderm, endoderm, and mesoderm.
- The digestive cavity has only one opening for both ingestion and egestion so food cannot be processed continuously.
- They have a solid body with no room for true digestive or respiratory systems to circulate food or oxygen. Flatworms have solved this problem in a unique way. The body is so flat and thin that many body cells can exchange nutrients and wastes by diffusion with the environment.

Nematodes—Roundworms

- Nematodes are unsegmented worms with bilateral symmetry but little sensory apparatus.
- Many are parasitic. *Trichinella* causes trichinosis, which is contracted by eating uncooked pork
- One species, *C. elegans,* is widely used as a model in studying the link between genes and development.

Annelids—Segmented Worms like Earthworms, Leeches

- Bilateral symmetry with little sensory apparatus
- Digestive tract is a tube-within-a-tube consisting of **crop**, **gizzard**, and intestine
- **Nephridia** for excretion of the nitrogen waste, urea
- **Closed circulatory system**—heart consists of five pairs of aortic arches
- Blood contains hemoglobin and carries oxygen
- Diffusion of oxygen and carbon dioxide through moist skin
- Hermaphrodites

Mollusks—Squids, Octopuses, Slugs, Clams, and Snails

- Have **soft body** often protected by a hard calcium-containing shell
- Have bilateral symmetry with three distinct body zones:

  1. Head-foot, which contains both sensory and motor organs
  2. Visceral mass, which contains the organs of digestion, excretion, and reproduction
  3. Mantle, a specialized tissue that surrounds the visceral mass and secretes the shell

- Radula, a movable, tooth-bearing structure, acts like a tongue
- Open circulatory system with blood-filled spaces called hemocoels or sinuses
- Most have gills and nephridia

Arthropods—Insecta (Grasshopper), Crustacea (Shrimp, Crab), Arachnida (Spider)

- Jointed appendages
- Segmented into head, thorax, abdomen
- More sensory apparatus than in annelids, giving them more speed and freedom of movement
- Chitinous exoskeleton protects the animal and aids in movement
- **Open circulatory system** with a tubular heart and **hemocoels**, sinuses
- **Malpighian tubules** for removal of nitrogenous wastes, uric acid
- Air ducts called trachea bring air from the environment into hemocoels

Echinoderms—Sea Stars (Starfish) and Sea Urchins

- Most are sessile or slow moving.
- They have bilateral symmetry as an embryo but revert to the primitive radial symmetry as an adult. The radial anatomy of the adult is an adaptation to a sedentary lifestyle.
- Their water vascular system creates hydrostatic support for the tube feet, the locomotive structures.
- Echinoderms reproduce by sexual reproduction with external fertilization.
- They can also reproduce by fragmentation and regeneration. Any piece of a sea star that contains part of the central canal will form a completely new organism.
- Sea stars have an endoskeleton consisting of calcium plates. An endoskeleton grows with the body. In contrast, an exoskeleton does not and must be shed periodically.

Chordates—Fish, Amphibians, Reptiles, Birds, Mammals

- Chordates have a **notochord**, a rod that extends the length of the body and serves as a flexible axis.
- They have a dorsal, hollow nerve cord.
- The tail aids in movement and balance. The coccyx bone in humans is a vestige of a tail.
- Birds and mammals are **homeotherms**—they maintain a consistent body temperature. All other chordates—fish, amphibians, and reptiles—are **poikilotherms** (cold-blooded) although some reptiles are endotherms (heat from within) and are able to raise their body temperature.

CHARACTERISTICS OF MAMMALS

- Mammals belong to the phylum Chordata.
- Mothers nourish their babies with milk from mammary glands.
- They have hair or fur.
- Mammals are homeotherms (warm-blooded).
- Most are placental mammals (eutherians)—the embryo develops internally in a uterus connected to the mother by a placenta, where nutrients diffuse from mother to embryo.
- Some, the marsupials, including kangaroos, are born very early in embryonic development. The "joey" completes its development while nursing in the mother's pouch attached to a teat.
- Monotremes, egg-laying mammals, like the duck-billed platypus and the spiny anteater, derive nutrients from a shelled egg.

Table 12.2 shows the classification of three mammals using our current system of taxonomy originally developed by Linnaeus.

HELPFUL HINT

Birds and mammals are both warm-blooded.

TABLE 12.2

Sample Classification

Taxa	Human	Lion	Dog
Domain	Eukarya	Eukarya	Eukarya
Kingdom	Animalia	Animalia	Animalia
Phylum	Chordata	Chordata	Chordata
Class	Mammalia	Mammalia	Mammalia
Order	Primate	Carnivora	Carnivora
Family	Hominid	Felidae	Canidae
Genus	*Homo*	*Panthera*	*Canis*
Species	*sapiens*	*leo*	*familiaris*

CHARACTERISTICS OF PRIMATES

Humans are primates. Primates descended from insectivores, probably from small, tree-dwelling mammals. Primates have dexterous hands and opposable thumbs, which make it possible to do fine-motor tasks. Nails have replaced claws. Hands and fingers contain many nerve endings and are sensitive. The eyes of a primate are front facing and set close together. Front-facing eyes fosters face-to-face communication. Close-set eyes are responsible for overlapping fields of vision, which enhance depth perception and hand-eye coordination. Although mammals devote much energy to the parenting of young, primates engage in the most intense parenting of any mammal. Primates usually have single births and nurture their young for a long time. Primates include humans, gorillas, chimpanzees, orangutans, gibbons, and the old world and new world monkeys.

MULTIPLE-CHOICE QUESTIONS

<u>Questions 1–9</u>

Refer to the terms below.

(A) Prokaryotes
(B) Fungi
(C) Plants
(D) Animals
(E) Protista

1. Contains the most diverse eukaryotic organisms

2. Heterotrophs whose cell walls consist of chitin

3. Included in the domain Bacteria

4. Heterotrophs that reproduce sexually with a dominant diploid stage

5. All are autotrophic eukaryotes

6. The Archaea domain includes these

7. The heterotrophic part of any lichen

8. Includes amoebas, paramecia, and euglenas

9. Eukaryotic cells that play a vital role in recycling nutrients in an ecosystem

10. All of the following are true of Protista EXCEPT

    (A) all are heterotrophs
    (B) all are eukaryotes
    (C) they include amoeba, paramecia, and euglena
    (D) some move by pseudopods, some by cilia, and some by flagella
    (E) they include the widest variety of organisms of any kingdom

11. Which of the following contains organisms capable of surviving extreme conditions of heat and salt concentration?

    (A) Archaea
    (B) Animalia
    (C) Protista
    (D) Fungi
    (E) Plants

<u>Questions 12–17</u>

Refer to this list of animals below.

    (A) Platyhelminthes
    (B) Nematodes
    (C) Chordates
    (D) Echinoderms
    (E) Cnidarians

12. Roundworms, all are parasites

13. Includes the flatworm planaria

14. Jellyfish and hydra

15. All organisms in the phyla contain stinging cells

16. Include mammals

17. Includes monotremes and marsupials

18. All of the following are characteristics of primates EXCEPT

    (A) opposable thumbs
    (B) front-facing eyes
    (C) nurture their young for a long time
    (D) examples are dolphins and whales
    (E) usually have single births

19. The ectoderm becomes the

    (A) nervous system
    (B) digestive organs
    (C) blood
    (D) bones
    (E) muscles

20. The endoderm becomes the

    (A) skin
    (B) digestive organs
    (C) blood
    (D) bone
    (E) muscle

EXPLANATION OF ANSWERS

1. **(E)** The Protista are all eukaryotes and include the widest variety of organisms. Some of these organisms are placed into this kingdom because they do not fit anywhere else.

2. **(B)** Fungi are all heterotrophic eukaryotes that carry out extracellular digestion. Their cell walls consist of chitin, not cellulose, as is found in the cell walls of plants.

3. **(A)** Prokaryotes are spread across two domains, Archaea and Bacteria. A prokaryote is a cell that has no nucleus or other internal membranes.

4. **(D)** Notice that the question specifies heterotrophs. Plants also have a dominant diploid stage, but they are all autotrophs. In animals, all the body cells are diploid ($2n$). Only the gametes, which cannot live on their own, are monoploid (n).

5. **(C)** Plants are autotrophic eukaryotes that have a dominant diploid stage. All the cells of a rose, for example, are diploid ($2n$). Only the gametes, which are dependent on the diploid cells and exist inside the male and female organs, are monoploid (n).

6. **(A)** The Archaea domain includes prokaryotes that live in the most extreme environments, such as hot springs.

7. **(B)** Lichens are symbiotic associations between a fungus and a photosynthetic organism, either a green algae or cyanobacteria. Lichens are extremely drought and cold resistant and can grow where few organisms can. They are often pioneer organisms, the first organisms to inhabit a barren environment.

8. **(E)** The Protista kingdom includes the most varied organisms. Some are autotrophs and some are heterotrophs, but all are eukaryotes.

9. **(B)** Fungi, along with the bacteria of decay, decompose organic matter and thus play an important role in ecosystems. Fungi are eukaryotes; bacteria are prokaryotes.

10. **(A)** Many Protista, like paramecia and amoebas, are heterotrophs. Some, like algae, are autotrophs. The euglena is an autotroph. The other choices are all true statements.

11. **(A)** The domain Archaea consists of prokaryotes that live in the most extreme and hostile environments, such as hot springs. In the former five-kingdom system of classification, they were included in the Monerans.

12. **(B)** The Nematodes includes the roundworms, unsegmented worms with bilateral symmetry but little sensory apparatus. Most are parasitic, like *Trichinella*, which comes from uncooked pork and causes trichinosis.

13. **(A)** The Platyhelminthes are the flatworms, which include tapeworms. These are the simplest animals with bilateral symmetry, an anterior end, and three distinct cell layers: the ectoderm, endoderm, and mesoderm.

14. **(E)** Jellyfish (which are not fish) and hydra are examples of cnidarians.

15. **(E)** The cnidarians all have stinging cells, cnidocytes containing stingers called nematocysts. They have only two cell layers: ectoderm and endoderm connected by noncellular mesoglea. They have no transport system; every cell is in direct contact with the environment.

16. **(C)** The chordates are characterized by having a notochord, a rod that extends the length of the body, which serves as a flexible axis. In this kingdom are the phyla of animals that are familiar to us: fish, amphibians, reptiles, and mammals.

17. **(C)** Monotremes (egg-laying mammals) and marsupials (mammals with a pouch, like kangaroos) are classified as chordates. Other chordates are fish, amphibians, and reptiles.

18. **(D)** Dolphins and whales are mammals, not primates. Primates are a subset of mammals. Examples of primates are humans, gorillas, chimpanzees, and orangutans. All the other choices are correct.

19. **(A)** The ectoderm becomes the skin and nervous system. The endoderm becomes the digestive organs. The mesoderm becomes the blood, bones, and muscles.

20. **(B)** The endoderm becomes the digestive organs. The ectoderm becomes the skin and nervous system. The mesoderm becomes the blood, bones, and muscles.

ORGANISMAL BIOLOGY

Plants

CHAPTER 13

> - Classification of plants
> - Strategies enabling plants to move to land
> - Primary and secondary growth
> - Roots—structure, function, and types
> - Stems—structure and function
> - Leaf—structure and function
> - Stomates
> - Types of plant tissue
> - Transport in plants
> - Plant reproduction—asexual and sexual
> - Structure of the seed
> - Alternation of generations
> - Plant responses to stimuli— hormones and tropisms

Plants include all multicelled, eukaryotic, photosynthetic **autotrophs**. Their cell walls are made of cellulose, and they store carbohydrates as starch. Biologists believe that modern, multicelled plants evolved from the green algae Chlorophyta.

CLASSIFICATION OF PLANTS

Plants can be classified as either bryophytes or tracheophytes.

Bryophytes

- Bryophates are primitive plants that lack vascular tissue.
- They must live in moist environments because they have no roots or xylem and must absorb water by diffusion.
- Bryophates are tiny because they lack the lignin-fortified tissue necessary to support tall plants on land.
- Mosses are an example.

Tracheophytes

- Tracheophytes have transport vessels, xylem and phloem.
- They include ancient seedless plants, like ferns, that reproduce by spores.
- They include modern plants that reproduce by seeds.
- Those with seeds are further subdivided into **gymnosperms** and **angiosperms**.

Gymnosperms

Gymnosperms are conifers, the cone-bearing plants. They have various modifications to help them survive under dry conditions. These include needle-shaped leaves, a thick and waxy cuticle, and stomates located in stomatal crypts to reduce water loss even further. Cedars, sequoias, redwoods, pines, yews, and junipers are all gymnosperms.

Angiosperms

Angiosperms are flowering plants. They are also called anthophyta. They are the most diverse and plentiful plants on Earth. Roses, daisies, fruits, nuts, grains, and grasses are just a few of the examples of angiosperms. These plants are subdivided into two more groups: **monocotyledons** (monocots) and **dicotyledons** (dicots). Table 13.1 shows the main differences between these two groups.

TABLE 13.1

The Principal Differences Between Monocots and Dicots		
Characteristic	**Monocots**	**Dicots**
Cotyledons (seed leaves)	One	Two
Vascular bundles in stem	Scattered	In a ring
Leaf venation	Parallel	Netlike
Floral parts	Usually in 3s	Usually in 4s or 5s
Roots	Fibrous roots	Taproots

Examples of monocots are the grasses: wheat, corn, oats, lawn grass, and rice. Monocots provide the food for most of the world. Palm trees are also monocots. Examples of dicots are daisies, roses, carrots, and most flowering plants you could think of. Oak, walnut, cherry, and most other trees you could name are dicots.

STRATEGIES THAT ENABLED PLANTS TO MOVE TO LAND

Plants began life in the seas and moved to land as competition for resources increased. The biggest problems a plant on land faces are supporting a plant body and absorbing and conserving water. Several modifications enable plants to live on land.

- Cell walls made of cellulose lend support to the plant whose cells, unsupported by a watery environment, must maintain their own shape.
- Roots and root hairs absorb water and nutrients from the soil.
- **Stomates** open to exchange photosynthetic gases and close to minimize excessive water loss.
- The waxy coating on the leaves, **cutin**, helps prevents excess water loss from the leaves.
- In some plants, gametes and zygotes form within a protective jacket of cells called **gametangia** that prevents drying out.

- **Sporopollenin**, a tough polymer, is resistant to almost all kinds of environmental damage and protects plants in a harsh terrestrial environment. It is found in the walls of spores and pollen.
- Seeds and pollen have a protective coat that prevents desiccation. They are also a means of dispersing offspring.
- Reduction of the primitive gametophyte (*n*) generation occurs.

HOW PLANTS GROW

Unlike animals, plants continue to grow as long as they live because plants have **meristem tissue** that continually divides, generating new cells. Plants grow in two ways: **primary growth** and **secondary growth.**

Primary Growth

Primary growth is vertical. It is the elongation of the plant down into the soil and up into the air. New cells arise from the constantly dividing growth layer called the apical meristem, which is located at the buds of shoots and the tips of the roots.

Root growth is concentrated near the root tip. Three zones of cells at different stages of primary growth are located there: the zone of cell division called apical meristem, the zone of elongation, and the zone of differentiation. The root tip is protected by a root cap that secretes a substance that helps digest the earth as the root tip grows through the soil.

Figure 13.1 shows a longitudinal section of a root. Be able to identify the different regions.

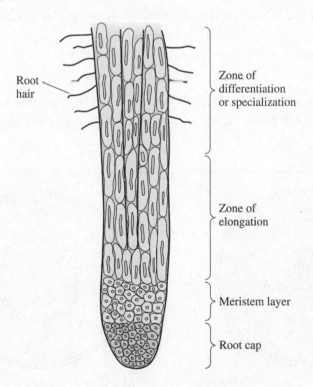

Figure 13.1 Root tip

ZONE OF CELL DIVISION

This zone contains meristem cells that are actively dividing and are responsible for producing new cells that grow down into the soil. This is the region you observed under the microscope in lab when you were studying cells undergoing mitosis.

ZONE OF ELONGATION

Cells in this zone elongate and are responsible for pushing the root cap downward and deeper into the soil.

ZONE OF DIFFERENTIATION

Cells in this zone undergo specialization into three primary meristems that give rise to three tissue systems in the plant: the epidermis, ground tissue, and xylem and phloem.

Secondary Growth

Secondary growth means lateral growth or an increase in girth. New cells are provided by the lateral meristem. In herbaceous (nonwoody) plants, like vegetables and flowers, there is only primary growth because these plants live for only one season. However, woody plants are protected by bark and live for many years. In these plants, secondary growth is responsible for the enlargement of the trunk. For each year of growth, another ring is added.

STUDY TIP

Be sure you can distinguish the monocot and dicot roots.

ROOTS

The function of the roots is to absorb nutrients from the soil, anchor the plant, and store food.

Structure

The root consists of specialized tissues and structures organized to carry out the various functions of the roots. Figure 13.2 is a sketch of cross sections of monocot and dicot roots.

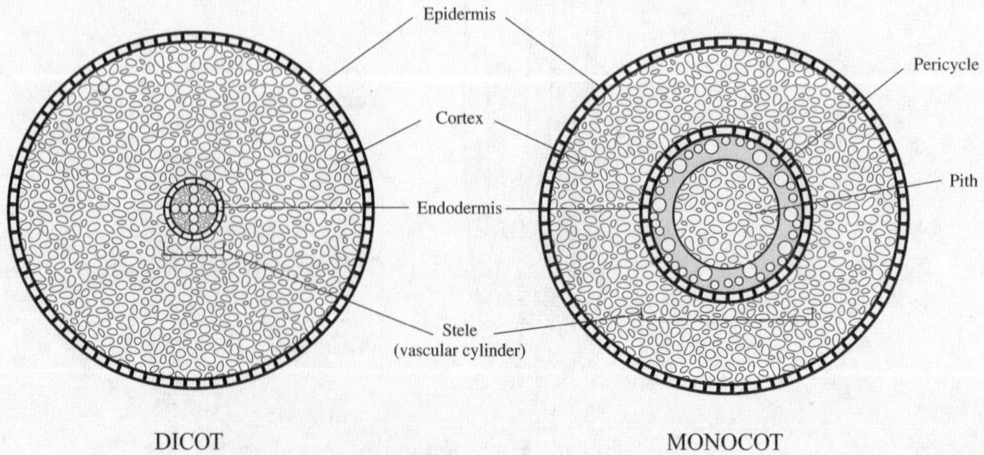

Figure 13.2 Roots

EPIDERMIS

The epidermis covers the entire surface of the root and is modified for absorption. Slender cytoplasmic projections from the epidermal cells, called root hairs, extend out from each cell and greatly increase the root's absorptive surface area.

CORTEX

The cortex consists of **parenchyma cells** that contain many **plastids** for the storage of starch and other organic substances.

STELE

The vascular cylinder or stele of the root consists of vascular tissues (xylem and phloem) surrounded by one or more layers of tissue called the pericycle, from which lateral roots arise.

ENDODERM

The vascular cylinder is surrounded by a tightly packed layer of cells called the endodermis. Each endoderm cell is wrapped with the Casparian strip, a continuous band of waxy material that is impervious to water and dissolved minerals. The function of the endoderm is to select what minerals enter the vascular cylinder and the body of the plant.

Types of Roots

The **taproot** is a single, large root that gives rise to lateral branch roots. In many dicots, the primary root is the taproot. Some taproots "tap" water deep in the soil. Others, like carrots, beets, and turnips, are modified for storage.

A fibrous roots system, common in monocots like grasses, holds the plant firmly in place. As a result, grasses make fine ground cover because they minimize soil erosion.

Adventitious roots are roots that arise above ground. Here are two examples:

1. **Aerial roots.** Trees that grow in swamps or salt marshes like mangroves have aerial roots that stick up out of the water and serve to aerate the root cells. English ivy has aerial roots that enable the ivy to cling to the sides of buildings.

2. **Prop roots.** Some tall plants like corn have prop roots that grow aboveground out from the base of the stem and help support the plant.

STEMS

The function of the stem is to support the plant. This support allows the leaves to receive the most light. Stems also transport water and minerals from the soil, and nutrients from the leaves to the rest of the plant.

Structure

Vascular tissue runs the length of the stem in strands called vascular bundles. Each bundle contains xylem on the inside, phloem on the outside, and meristem tissue between the two. In monocots, the vascular bundles are scattered throughout the stem. In dicots, they are arranged in a ring around the edge of the stem. The ground tissue of the stem consists of cortex and **pith**, parenchymal tissues modified for storage.

Figure 13.3 shows the difference between monocot and dicot stems.

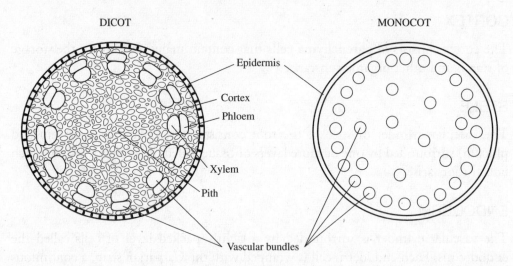

Figure 13.3 Stems

THE LEAF

The leaf is organized to maximize sugar production while minimizing water loss. Table 13.2 lists the parts of a leaf and their functions, and Figure 13.4 illustrates them.

TABLE 13.2

Leaf Parts and Function	
Parts of Leaf	**Function**
Epidermis—upper and lower	Protection
Waxy cuticle—made of cutin	Minimizes water loss
Guard cells—modified epidermal cells, contain chloroplasts	Control the opening of the stomates
Palisade mesophyll—tightly packed	Photosynthesis
Spongy mesophyll—loosely packed	Photosynthesis
	Diffusion and exchange of gases into and out of these cells
Veins—located in the mesophyll	Carry water and nutrients from the soil to the leaves and carry sugar, the product of photosynthesis, from the leaves to the rest of the plant.

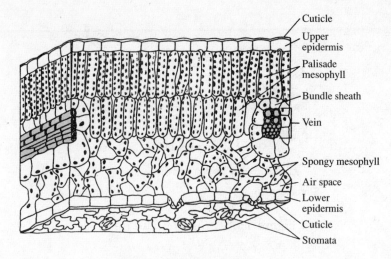

Figure 13.4 The leaf

Stomates

When plant cells use the food they make, they carry out cellular respiration, just like animals do. This means they take in oxygen and give off carbon dioxide. When plant cells carry out photosynthesis, however, they take in carbon dioxide and give off oxygen and water vapor. Plants exchange these gases between air spaces in the spongy mesophyll and the exterior of the leaf by opening their stomates. So why do plants ever close their stomates? If stomates were kept open all the time, the plant would lose so much water through **transpiration** (loss of water from the leaf) it could not survive. To minimize excessive water loss, when the sun is shining brightly and photosynthesis is running at top speed, stomates are open. At night, though, most plants close their stomates.

> **INTERESTING FACT**
>
> Stomates are a necessary evil! Read why.

Plants must keep their stomates open enough to allow photosynthesis to take place but not so much that they lose an excess amount of water.

Guard cells are modified epithelial cells containing chloroplasts that control the opening and closing of the stomates by changing their shape. The cell walls of guard cells are not uniformly thick. Cellulose microfibrils are oriented in such a direction (radially) that when the guard cells absorb water by osmosis and become **turgid**, they curve like hot dogs, causing the stomate to open. When guard cells lose water and become flaccid, the stomate closes. Figure 13.5 is a sketch of open and closed guard cells and stomates in a leaf.

FACTOR THAT CAUSES STOMATES TO OPEN

Depletion of carbon dioxide within the air spaces of the leaf causes stomates to open.

FACTORS THAT CAUSE STOMATES TO CLOSE

- Lack of water causes the guard cells to lose their turgor, become flaccid, and close the stomates.
- High temperatures also close the stomates presumably by stimulating cellular respiration and increasing carbon dioxide concentration within the air spaces of the leaf.

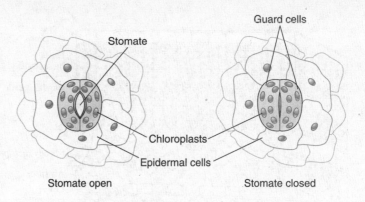

Figure 13.5 Stomates and guard cells

TYPES OF PLANT TISSUE

Just as there are different cell and tissue types in animals, plants too have different cell and tissue types. Plants consist of three main tissue types: dermal, vascular, and ground tissue. A fourth tissue type, meristem tissue or growth tissue, is found only in the growing tips of shoots and roots. It is also discussed in the section, "How Plants Grow."

Dermal Tissue

Dermal tissue is the outer protective covering of plants and usually consists of a single layer of epidermal cells. On leaves, epidermal cells are protected by the cuticle, which is made of the waxy molecule cutin. Some leaves are also covered with tiny, spikelike projections called **trichomes**, which also protect the leaf. For the most part, epidermal cells do not contain chloroplasts and cannot photosynthesize. An important exception are guard cells, which are modified epidermal cells that contain chloroplasts and can photosynthesize.

Vascular Tissue

Vascular tissue transports water and nutrients up and down the plant. There are two types: xylem and phloem. Xylem consists of **tracheids** and **vessel elements**. Phloem consists of **sieve tube elements** and **companion cells**.

Ground Tissue

Ground tissue makes up all plant tissue besides dermal and vascular tissue. It consists of three cell types: parenchyma, collenchyma, and sclerenchyma.

PARENCHYMA CELLS

Parenchyma cells are the traditional-looking plant cell. They have a primary cell wall that is thin and flexible, and they lack a secondary cell wall. The cytoplasm contains one or two large vacuoles. When the cell is turgid (swollen) with water, these cells lend support to the plant. They are found in all parts of the plant. Some, like mesophyll cells, contain chloroplasts. Others, like epidermal cells, do not.

COLLENCHYMA CELLS

celery

Collenchyma cells have unevenly thickened primary cell walls but lack secondary cell walls. The "strings" of celery consist of collenchyma cells.

SCLERENCHYMA CELLS

Sclerenchyma cells have very thick primary and secondary cell walls that are fortified with lignin. Their function is purely for support.

TRANSPORT IN PLANTS

Just like animals, plants need to transport water, nutrients, and gases. Unlike animals, plants do not have blood, arteries, or a heart to accomplish this. Instead, they have xylem and phloem.

Xylem

Xylem consists of two types of elongated cells: tracheids and vessel elements. The secondary cell walls of tracheids are hardened with lignin and function to support the plant as well as to transport nutrients and water. Xylem is what makes up the stuff we call wood.

Xylem carries water and nutrients from the soil up to the tallest leaves against gravity with *no expenditure of energy*. Instead, they are pulled up by a combination of two phenomena: transpirational pull and cohesion tension. Transpiration is the evaporation of water from leaves. Cohesion refers to the fact that water molecules are attracted to each other and stick together. The transpirational pull-cohesion tension theory states that for each molecule of water that evaporates from a leaf by transpiration, another molecule of water is drawn in at the root to replace it. The absorption of sunlight drives transpiration by causing water to evaporate from the leaf. Several factors affect the rate of transpiration and loss of water from a leaf.

- High humidity slows down transpiration, while low humidity speeds it up.
- Wind can reduce humidity near the stomates and thereby increase transpiration.
- Increased light intensity will increase photosynthesis, thereby increasing both the amount of water vapor to be transpired and the rate of transpiration.
- Closing stomates stops transpiration.

Phloem

Phloem vessels are made of chains of two types of cells: sieve tube elements and companion cells. They carry sugar from the photosynthetic leaves to the rest of the plant by a process called **translocation**. Sugar is stored in the roots. Unlike transport in the xylem, *this process requires energy*.

ABSORPTION OF NUTRIENTS AND WATER

Plants use their roots to absorb nutrients and water from the soil. These then must be absorbed by the cells themselves.

Apoplast and Symplast

The movement of water and solutes across a plant, called lateral movement, is accomplished along the symplast and apoplast. The **symplast** is a continuous system of cytoplasm of cells interconnected by **plasmodesmata**. The apoplast is the network of cell walls and intercellular spaces within a plant body that permits extensive extracellular movement of water within a plant.

Mycorrhizae

In mature plants of many species where older regions of roots lack root hairs, **mycorrhizae** supply the plant with water and minerals. Mycorrhizae are symbiotic structures consisting of the plant's roots intermingled with the hyphae (filaments) of a fungus that greatly increase the quantity of nutrients that a plant can absorb.

Rhizobium

Rhizobium is a symbiotic bacterium that lives in the nodules on roots of specific legumes. It fixes nitrogen gas from the air into a form of nitrogen the plant requires.

PLANT REPRODUCTION

Plants can reproduce both asexually and sexually.

Asexual Reproduction

Plants can clone themselves or reproduce asexually by **vegetative propagation**. In this process, a piece of the vegetative part of a plant, the root, stem, or leaf, produces an entirely new plant genetically identical to the parent plant. Examples are grafting, cuttings, bulbs, and runners.

Sexual Reproduction in Flowering Plants

The flower is the sexual organ of a plant. Figure 13.6 shows the structure and function of the parts of the flower.

1. **Petals.** Brightly colored, modified leaves found just inside the circle of sepals; attract animals that will pollinate the plant

2. **Sepals.** Outermost circle of leaves; are green and closely resemble ordinary leaves; enclose the bud before it opens and protects the flower while it develops

3. **Pistils or carpels.** Female part of the flower; produce the female gameto-phytes; each consists of an ovary, stigma, and style

4. **Ovary.** Swollen part of pistil that contains the ovule, where one or more ova are produced

5. **Ovule.** The structure within the ovary where the ova (female gametophytes) are produced

6. **Style.** Long, usually thin stalk of the pistil

7. **Stigma.** Sticky top of the style where pollen lands and germinates

8. **Stamen.** Male part of the flower, made up of anther and filament

9. **Anther.** Male part of the flower where sperm (pollen) are produced by meiosis

10. **Filament.** Threadlike structure that supports the anther

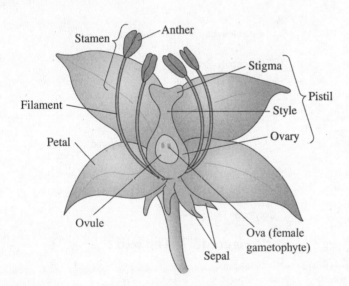

Figure 13.6 The flower

POLLINATION AND FERTILIZATION

Sexual reproduction begins with pollination. One pollen grain containing three monoploid nuclei—one tube nucleus and two sperm nuclei—lands on the sticky stigma of the flower. The pollen grain absorbs moisture and germinates or sprouts, producing a pollen tube that burrows down the style into the ovary. The two sperm nuclei travel down the pollen tube into the ovary. Once inside the ovary, the two sperm nuclei enter the ovule through the micropyle. One sperm nucleus fertilizes the egg and becomes the embryo (2n). The other sperm nucleus fertilizes the two polar bodies and becomes the triploid (3n) endosperm, the food for the growing embryo.

This process is known as double fertilization because two fertilizations occur. After fertilization, the ovule becomes the seed and the ripened ovary becomes the fruit. In monocots, food reserves remain in the endosperm. In dicots, the food reserves of the endosperm are transported to the cotyledons, and consequently, the mature dicot seed lacks endosperm. In the monocot the coconut, the endosperm is liquid.

Double fertilization can be seen as:

1. Sperm + Ovum $\rightarrow$ Embryo = $2n$
2. Sperm + 2 Polar bodies $\rightarrow$ Cotyledon (Food for the growing embryo) = $3n$

THE SEED

The seed consists of a protective seed coat, an embryo, and the cotyledon or endosperm—food for the growing embryo. The embryo consists of the **hypocotyl**, **epicotyl**, and **radicle**. The **hypocotyl** becomes the lower part of the stem and the roots. The **epicotyl** becomes the upper part of the stem. The radicle, or embryonic root, is the first organ to emerge from the germinating seed.

Figure 13.7 shows a dicot seed (like a peanut) split in half. A monocot seed, like corn, does not split in half. In addition, the food source in a monocot is endosperm instead of cotyledon.

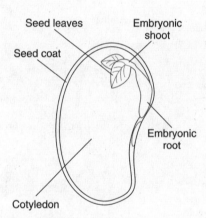

Figure 13.7 The seed

REMEMBER
The gametophyte generation is monoploid (n).
The sporophyte generation is diploid ($2n$).

ALTERNATION OF GENERATIONS

The sexual life cycle of plants is characterized by the alternation of generations in which monoploid (n) and diploid ($2n$) generations alternate with each other. The **gametophyte** (n) produces gametes by mitosis that fuse during fertilization to yield $2n$ zygotes. Each zygote develops into a **sporophyte** ($2n$) that produces monoploid spores (n) by meiosis. Each monoploid spore forms a new gametophyte, and the cycle continues, as seen in Figure 13.8.

This topic is complex, but questions on it are usually basic on standardized tests. You should study the vocabulary listed in Table 13.3 and have a general sense of the topic.

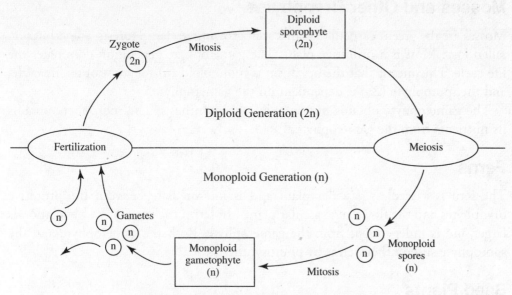

Figure 13.8 Alternation of generations

TABLE 13.3

Vocabulary for Alternation of Generations

Term	Definition
Antheridium	Structure that produces sperm, develops on the gametophyte
Archegonium	Structure that produces eggs, develops on the gametophyte
Gametophyte	A monoploid adult plant
Megaspores	Produced by large female cones and will develop into female gametophytes
Microspores	Produced by small male cones and will develop into male gametophytes or pollen grains
Protonema	Branching, one-celled-thick filaments produced by germinating moss spores, becomes the gametophyte in moss
Sporangia	Located on the tip of the mature sporophyte, where meiosis occurs, producing monoploid spores
Sporophyte	A diploid adult plant
Sori	Raised spots located on the underside of sporophyte ferns, clusters of sporangia

Mosses and Other Bryophytes

Mosses are the green, carpetlike plants seen growing in damp forests, sometimes on fallen logs. Moss is a primitive plant. The gametophyte generation dominates the life cycle. This means that the organism is monoploid (n) for most of its life cycle, and the sporophyte ($2n$) is dependent on the gametophyte.

The gametophyte obtains nutrients by photosynthesis. The sporophyte obtains its nutrients from the gametophyte.

Ferns

The fern is a seedless vascular plant and is intermediate between the primitive bryophytes and the flowering vascular plants. In ferns, the sporophyte generation is larger and is independent from the gametophyte. Both the gametophyte and the sporophyte sustain themselves by photosynthesis.

Seed Plants

Seed plants are advanced, vascular plants. They are divided into two groups: flowering plants and conifers. In the flowering plants (angiosperms), the gametophyte generation exists inside the sporophyte generation and is totally dependent on the sporophyte.

In a gymnosperm (cone-bearing plant) like the pine tree, the gametophyte generation develops from haploid spores that are retained within the sporangia.

PLANT RESPONSES TO STIMULI

Although plants cannot run away from stimuli, they can still react to them.

Hormones

Plant hormones help coordinate growth, development, and response to environmental stimuli. They are produced in very small quantities. However, they have a profound effect on the plant because the hormone signal is amplified. A plant's response to a hormone usually depends not so much on absolute quantities of hormones but on relative amounts. Hormones can have multiple effects on a plant, and they can work synergistically with other hormones or in opposition to them. The following is an overview of plant hormones and what they stimulate.

AUXINS

- Phototropisms occur due to an unequal distribution of auxins.
- Auxins enhance apical dominance, the preferential growth of a plant upward (toward the sun) rather than laterally. The terminal bud actually suppresses lateral growth by suppressing development of axial buds.
- Auxins stimulate stem elongation and growth by softening the cell wall.
- The first plant hormone discovered was auxin.
- Indoleacetic acid (IAA) is a naturally occurring auxin.
- A human-made auxin, 2,4-D, is used as a weed killer.

- Auxins are used as rooting powder to develop roots quickly in a plant cutting.
- A synthetic auxin sprayed on tomato plants will induce fruit production without pollination. This results in seedless tomatoes.

CYTOKININS

- **Cytokinins** stimulate cytokinesis and cell division.
- They work in concert with auxins to promote growth and cell division.
- Cytokinins work antagonistically against auxins in relation to apical dominance.
- They delay senescence (aging) by inhibiting protein breakdown. (Florists spray cut flowers with cytokinins to keep them fresh.)
- Cytokinins are produced in the roots and travel upward in the plant.

GIBBERELLINS

- **Gibberellins** promote stem and leaf elongation.
- They work in concert with auxins to promote cell growth.
- They induce bolting, the rapid growth of a floral stalk. When a plant, such as broccoli, which normally grows close to the ground, enters the reproductive stage, it sends up a very tall shoot on which the flower and fruit develop. This is a mechanism to ensure pollination and seed dispersal.

ABSCISIC ACID (ABA)

- **Abscisic acid** inhibits growth.
- It enables plants to withstand drought.
- It closes stomates during times of water stress.
- ABA works in opposition to the growth-promoting plant hormones.
- ABA counteracts the breaking of dormancy during a winter thaw.
- It promotes seed dormancy. This prevents seeds that have fallen onto the ground in the fall from sprouting until the spring when environmental conditions are better. In some desert plants, seeds break dormancy only after heavy rains have washed all ABA out of the cells. (The ground is moist and can support plant growth.)

ETHYLENE

- This plant hormone is a gas.
- **Ethylene** gas promotes ripening, which in turn, triggers increased production of ethylene gas. "One bad apple spoils the whole barrel." This is an example of positive feedback.
- Commercial fruit sellers pick perishable fruit before they are ripe, while still hard. When the fruit arrive at their destination, they are sprayed with ethylene gas to hasten the ripening. In contrast, apples are kept in an environment of CO_2 to eliminate exposure to ethylene gas and thus keep the apples from ripening or rotting. In this way, apples can be stored for long periods of time.
- Ethylene facilitates **apoptosis**, programmed cell death. Prior to death, cells break down many of their chemical components for the plant to salvage and reuse.
- Ethylene promotes leaf abscission, the leaf dies and falls from the plant. Winds common in the fall help to blow the leaves from their attachment on trees.

Tropisms

A **tropism** is the growth of a plant toward or away from a stimulus. Examples are thigmotropisms (touch), geotropisms or gravitropisms (gravity), and phototropisms (light). A growth of a plant toward a stimulus is known as a positive tropism, while a growth away from a stimulus is a negative tropism.

Phototropisms result from an unequal distribution of **auxins** that accumulate on the side of the plant away from the light. Since auxins cause growth, the cells on the shady side of the plant enlarge and the stem bends toward the light.

Geotropisms result from an interaction of auxins and statoliths, specialized plastids containing dense starch grains

MULTIPLE-CHOICE QUESTIONS

1. All of the following are correct about plants EXCEPT

    (A) gymnosperms have cones
    (B) tracheophytes have xylem and phloem
    (C) moss is a bryophyte
    (D) bryophytes must live in a moist environment
    (E) some plants have cell walls made of chitin instead of cellulose

Questions 2–7

Choose from the terms below.

    (A) Monocot
    (B) Dicot

2. Vascular bundles scattered throughout the stem

3. Parallel veins in leaves

4. Seed splits in two

5. Taproots

6. Daisies, carrots, and roses are examples

7. Grasses such as lawn grass, wheat, rice, and corn are examples

8. All of the following are modifications for a dry environment EXCEPT

   (A) stomatal crypts
   (B) sporopollenin
   (C) seeds
   (D) flowers
   (E) pollen

9. Angiosperms are

   (A) monocots
   (B) seedless plants
   (C) plants with no vascular tissue
   (D) flowering plants
   (E) plants that reproduce only asexually

10. All of the following pairs are matched correctly EXCEPT

   (A) gymnosperm—cone bearing
   (B) ferns—seedless
   (C) flowering plants—reproduce by spores
   (D) bryophytes—moss
   (E) dicot—peanut

11. One pollen grain contains

   (A) one monoploid nucleus
   (B) two monoploid nuclei
   (C) three monoploid nuclei
   (D) one diploid nucleus
   (E) two diploid nuclei

12. The hypocotyl becomes the

   (A) seed coat
   (B) anther
   (C) sepals
   (D) embryo
   (E) lower part of the growing plant

13. Where are sperm produced in a flowering plant?

   (A) Filament
   (B) Anther
   (C) Epicotyl
   (D) Pistil
   (E) Carpel

14. Which of the following is NOT an example of asexual reproduction?

   (A) Runners
   (B) Grafting
   (C) Seeds
   (D) Cuttings
   (E) Bulbs

15. All of the following are correct about bryophytes EXCEPT

   (A) there is no vascular tissue
   (B) they are so hardy that they can be found anywhere in nature
   (C) the gametophyte is dominant
   (D) archegonia produce eggs
   (E) an example is moss

16. All of the following are correct about alternation of generations EXCEPT

   (A) in ferns, the sporophyte is dominant
   (B) the sporophyte is diploid
   (C) in flowering plants, the gametophyte is dominant
   (D) the gametophyte is monoploid
   (E) like all plants, bryophytes show alternations of generations

Questions 17–23

Choose from the terms below.

   (A) Auxins
   (B) Gibberellins
   (C) Abscisic acid
   (D) Ethylene
   (E) Cytokinins

17. Closes stomates during times of stress, inhibits growth

18. Is a gas

19. Enhances apical dominance

20. Induces bolting, the rapid growth of a floral stalk

21. An example is 2,4-D, a weed killer

22. Rooting powder is made of this

23. Promotes the ripening of fruit

24. Tropisms are caused by

    (A) ethylene
    (B) stimulation from primitive nerve cells
    (C) contraction of primitive muscle cells
    (D) unequal distribution of abscisic acid
    (E) unequal distribution of auxins

25. Statoliths are most closely related to

    (A) cone formation in gymnosperms
    (B) double fertilization
    (C) geotropisms
    (D) modified stomata
    (E) symbionts that live in nodules on the roots of plants

EXPLANATION OF ANSWERS

1. **(E)** All plants have cell walls made of cellulose. Fungi have cells walls made of chitin. All the rest are correct statements.

2. **(A)** Vascular bundles are scattered throughout the stems of monocots and organized in a ring around the edge in dicots.

3. **(A)** Veins are parallel in the leaves of monocots and netlike in the leaves of dicots.

4. **(B)** The cotyledon is the food for the growing embryo within the seed. Dicot is short for dicotyledon, which means the seed consists of two parts, or dicots. An example is a peanut. In monocots or monocotyledons, the seed consists of one part. An example is corn.

5. **(B)** A taproot is a single, large root that gives rise to lateral branch roots. In many dicots, the primary root is the taproot. Examples are carrots, turnips, and radishes. Monocots have fibrous roots.

6. **(B)** Daisies, carrots, and roses are all dicots. The grasses are all monocots. Examples of grasses are wheat, corn, hay, lawn grass, and rice. Monocots feed the world.

7. **(A)** The grasses are all monocots. Examples of grasses are wheat, corn, hay, lawn grass, and rice. Monocots feed the world. Daisies, carrots, and roses are all dicots.

8. **(D)** Flowers are organs for sexual reproduction, not specifically a modification for a dry environment.

9. **(D)** Angiosperms are flowering plants. Plants with no vascular tissue are called bryophytes. Flowers are an adaptation for sexual reproduction. Angiosperms are divided into two types: monocots and dicots.

10. **(C)** Flowering plants reproduce by seeds. Ferns reproduce by spores. All the other pairs are correctly matched.

11. **(C)** One pollen grain contains three monoploid nuclei: one tube nucleus and two sperm nuclei. The pollen grain sprouts and produces a pollen tube that

burrows down the style into the ovary. The two sperm nuclei travel down the pollen tube and into the ovary.

12. **(E)** The hypocotyl and epicotyl are parts of the embryo. The hypocotyl becomes the lower part of the stem and the roots. The epicotyl becomes the upper part of the stem and leaves.

13. **(B)** Meiosis occurs in the male part of a flowering plant, the stamen, more specifically within the anther. The filament supports the anther. The epicotyl is part of the embryo and becomes the upper part of the new plant. The pistil and carpel are synonyms for the female part of the flower.

14. **(C)** Seeds are the product of sexual fertilization. Seeds contain the embryo and food for the growing embryo, the cotyledon or endosperm.

15. **(B)** Bryophytes have no vascular tissue and absorb water by diffusion. That is why they must live in a moist environment. They often populate forests where it is cool and damp. They are very low to the ground because they have no vascular tissue (xylem and phloem). The gametophyte is dominant, and the sporophyte is dependent on the gametophyte. The archegonia produce eggs, and the antheridia produce sperm. Moss is an example of a bryophyte.

16. **(C)** In bryophytes the gametophyte (n) is dominant and the sporophyte is dependent on the gametophyte. In flowering plants, the opposite is true: the sporophyte is dominant. The other statements are all correct about alternation of generations.

17. **(C)** Abscisic acid (ABA) inhibits growth and enables plants to withstand drought by closing stomates.

18. **(D)** Ethylene is the only auxin that is a gas. It promotes the ripening of fruit.

19. **(A)** Auxins stimulate growth. They are used as rooting powder to stimulate rapid growth on a plant cutting.

20. **(B)** Gibberellins promote stem and leaf elongation. They induce bolting, the rapid growth of a floral stalk.

21. **(A)** Auxins stimulate growth. As odd as it sounds, they are also used as weed killers. 2,4-D is one weed killer. The auxin stimulates such rapid growth in the weed that it cannot take in adequate nutrients and it dies.

22. **(A)** Auxins stimulate growth. They are used as rooting powder to stimulate rapid growth on a plant cutting.

23. **(D)** Ethylene promotes the ripening of fruit. It is a gas.

24. **(E)** A tropism is the growth of a plant toward or away from a stimulus. They result from an unequal distribution of auxins in the upper stem.

25. **(C)** Statoliths are specialized plastids containing dense starch grains. They help the plant orient itself in space and to "know" where geologic center is.

Animal Physiology

CHAPTER 14

- Movement and locomotion
- Body temperature regulation
- Excretion
- Hydra—Phylum Cnidaria
- Earthworm—Phylum Annelida
- Grasshopper—Phylum Arthropods

This chapter offers a review of nonhuman animals and will help you gain a broader understanding into adaptations common to all animals. Although few questions from this area appear regularly in the SAT Subject Test, the animals discussed are commonly studied in many basic biology courses.

Animals are multicellular eukaryotes. All are **heterotrophs** and acquire nutrients by **ingestion**. This chapter reviews general concepts in the areas of movement and locomotion, body temperature regulation, and excretion. This is followed by specific information about three representative animals in three different phyla: hydra (Cnidaria), earthworm (Annelida), and grasshopper (Arthropod).

MOVEMENT AND LOCOMOTION

Movement is a characteristic of all animals. Movement can refer to the beating of cilia or the waving of tentacles to capture prey. It can also mean **locomotion**, movement from place to place. Some animals like hydra or sponges are **sessile**, meaning that they do not move. The hydra (phylum Cnidaria) feeds by moving its tentacles and stinging prey that swims near enough to touch.

Most animals spend their time and energy capturing food, seeking a mate, or escaping danger. Some mollusks, like clams, have a mantle, which secretes a shell that offers protection. Some Arthropods, like crabs and grasshoppers, have an exoskeleton consisting mostly of the polysaccharide, chitin, which does not grow with the animal and must be shed periodically. Not only does this exoskeleton protect the soft body inside, but in combination with muscles, it enables the animal to move rapidly. Animals, like the nematodes (roundworms), flatworms (planaria), and annelids (earthworms) have a hydrostatic skeleton, a closed body compartment filled with fluid. Muscles change the shape of this fluid-filled compartment, enabling the animal to move from place to place. Chordates, like frogs, cats, and humans, have an endoskeleton made of bone and cartilage that grows as the animal grows. Bones are connected to each other at joints by **ligaments**, while **tendons** connect bones to muscles.

> **REMEMBER**
>
> An endoskeleton grows with the animal; the exoskeleton does not.

BODY TEMPERATURE REGULATION

Most life exists only within a fairly narrow range, from 0°C, the temperature at which water freezes, to about 50°C. Animals must either seek out or create a suitable environment for themselves. The oceans are the most stable environment and experience the least fluctuation in environmental temperatures.

Temperatures on land, however, fluctuate enormously. Therefore, temperature regulation, like water conservation, became a problem for animals when they moved to land millions of years ago.

For example, the size of the ears in a jackrabbit can be correlated to the climate it lives in. Jackrabbits that evolved in cold, northern regions have small ears close to the head to minimize heat loss. Rabbits that evolved in warm, southern regions have long ears to dissipate heat from the many capillaries that make their ears appear pink.

Animals can also regulate body temperatures by changes in behavior. Here are some examples.

- A snake can warm itself in the sun and cool off by hiding in the shade.
- Animals on a cold prairie in winter huddle to decrease heat loss.
- Bees swarming in a hive raise the temperature inside the hive.
- Dogs pant and sweat through their tongues.
- Elephants lack sweat glands, but they wet down their thick skins with water and flap their ears, which are rich in capillaries.
- Humans shiver and jump around to keep warm.

Clarifying Terms

1. Cold-blooded and warm-blooded are not scientific terms.
2. Ectotherm means heated from outside and is probably closest in meaning to cold-blooded.
3. Endotherm (homeotherm) is the scientific word for warm-blooded. It means maintaining a constant body temperature despite fluctuations in the environmental temperature. Among animals, only birds and mammals are homeotherms. Although being warm-blooded requires enormous energy, it may have given birds and mammals an edge in ancient Earth when the dominant animal was the reptile. Mammals and birds can be active at any time, while the reptiles can be active only when the temperature permits it.

EXCRETION

Excretion is the removal of metabolic wastes. These include water, carbon dioxide, and nitrogenous wastes. There are three different types of nitrogenous wastes.

Ammonia

- Very soluble in water and highly toxic
- Excreted generally by organisms that live in water, including hydra and fish

STUDY TIP

Know these nitgoenous wastes:
- Ammonia
- Urea
- Uric acid

Urea

- Not as toxic as ammonia
- Excreted by earthworms and humans
- In mammals, is formed in the liver from ammonia

Uric Acid

- Pastelike substance that is not soluble in water and therefore not very toxic
- Excreted by insects, many reptiles, and birds, with a minimum of water loss

Different excretory mechanisms have evolved in various organisms for the purpose of removal of metabolic wastes. Table 14.1 lists some organisms matched with their structures.

TABLE 14.1

Excretion in Various Animals

Organism	Structures	Nitrogenous Waste
Protista	Contractile Vacuole	Ammonia
Platyhelminthes (planaria)	Flame cells	Ammonia
Earthworms	Nephridia (metanephridia)	Urea
Insects	Malpighian tubules	Uric acid
Humans	Nephrons	Urea

HYDRA—PHYLUM CNIDARIA

Nutrition

In Cnidarians like hydra and jellyfish, digestion occurs in the **gastrovascular cavity**, which has only one opening, the mouth. The animal has a two-way digestive tract, which means that food enters the same opening as waste exits. Cells of the gastrodermis (lining of the gastrovascular cavity or gastrocoel) secrete digestive enzymes into the cavity to aid in extracellular digestion where the main part of digestion occurs. Since the cnidarians are animals, their cells contain **lysosomes** that carry out intracellular digestion as well as extracellular digestion.

Body Plan and Symmetry

The basic body plan of the hydra is a **polyp**; while the body plan of the jellyfish is the **medusa**. The symmetry of all animals in this phylum is primitive and radial. The animal has only two cell layers, **ectoderm** and **endoderm**. The layers are held together by a middle layer called the **mesoglea** (middle glue). Every cell is in direct contact with its environment and the hydra has no need of a circulatory system.

Nervous System

All cnidarians have unique cells called **cnidocytes** that contain stingers, called **nematocysts**. Response to the environment is controlled by a primitive nervous system, a nerve net, where impulses travel in all directions from any site. As a result, the entire animal responds to a single stimulus. For example, if you knock the dish a hydra is in, its entire body will respond by shrinking into a tiny ball.

Reproduction

Cnidarians reproduce sexually, as well as asexually, by **budding**.

EARTHWORM—PHYLUM ANNELIDA

Nutrition

The digestive tract of the earthworm is a long, straight tube. As the earthworm burrows in the ground, creating tunnels that aerate the soil, the mouth ingests decaying organic matter along with soil. From the mouth, food moves to the esophagus and then to the **crop**, where it is stored. Posterior to the crop, the **gizzard**, which consists of thick muscular walls, grinds up the food with the help of sand and soil, which were ingested along with the organic matter. The rest of the digestive tract consists of the intestines where chemical digestion and absorption occur. Absorption is enhanced by the presence of a large fold in the upper surface of the intestine, called the typhlosole, which greatly increases the surface area.

Nervous and Transport Systems

In the earthworm, the exchange of respiratory gases—oxygen and carbon dioxide—between the environment and cells occurs passively by **diffusion** through moist skin. Earthworms are said to have an external respiratory surface because diffusion of these gases occurs at the animal's surface. The heart consists of five pairs of aortic arches that pump blood through the body in arteries, veins, and capillaries. Since blood never normally leaves these blood vessels, the earthworm has a closed circulatory system. Oxygen is carried by hemoglobin dissolved in red blood.

The brain of the earthworm consists of two dorsal, solid, fused ganglia that connect to a solid, ventral nerve cord.

Excretion

The earthworm has paired **nephridia** in every body segment to remove the nitrogenous waste urea.

Reproduction

The earthworm is a **hermaphrodite**, meaning it has both male and female sex organs.

GRASSHOPPER—PHYLUM ARTHROPODA

Nutrition

Like the earthworm, the grasshopper has a digestive tract that consists of a long tube consisting of a crop and gizzard. However, there are several differences between the two animals. The grasshopper has specialized mouthparts for tasting, biting, and crushing food and a gizzard that contains plates made of chitin that help grind food. In addition, the digestive tract is also responsible for removing the nitrogenous waste uric acid from the animal. **Malpighian tubules** serve this function.

Nervous and Transport Systems

The nervous system of the grasshopper is similar to that of the earthworm, but the transport system is different. The grasshopper heart is tubular, and the animal lacks capillaries. The grasshopper has an open circulatory system where blood normally leaves the artery and moves through interconnected **sinuses** or **hemocoels**, spaces surrounding the organs. Blood does not carry hemoglobin or oxygen.

Exchange of Respiratory Gases

The grasshopper and other arthropods and crustaceans have an internal respiratory surface because exchange of oxygen and carbon dioxide occurs inside the animal. Air enters the body through spiracles and travels through a system of tracheal tubes into the hemocoels or sinuses, where diffusion occurs. In arthropods and in some mollusks, oxygen is carried by hemocyanin, a molecule similar to hemoglobin but with copper, instead of iron, as its core atom.

MULTIPLE-CHOICE QUESTIONS

<u>Questions 1–13</u>

    (A) Hydra
    (B) Grasshopper
    (C) Earthworm

1. The body plan is a polyp

2. Has an open circulatory system

3. Has a closed circulatory system

4. Has no circulatory system

5. Nervous system consists of a nerve net

6. Exchange of respiratory gases is through the moist skin

7. Uses its digestive tract to release nitrogenous wastes

8. Uses nephridia to remove nitrogenous wastes

9. All the members of this phylum have stinging cells

10. This animal has a transport system but lacks capillaries

11. Every cell in this animal is directly in contact with its environment

12. Exchanges respiratory gases through hemocoels

13. Sessile

14. All of the following are behavioral changes to regulate body temperature EXCEPT

    (A) shivering
    (B) elephants bathing and spraying water on themselves
    (C) animals huddling at night on the prairie
    (D) increasing blood flow to the fingers and toes
    (E) resting on a hot rock

15. Which of the following correctly identifies an advantage of an internal skeleton over an external one?

    (A) An internal skeleton supports the animal better.
    (B) An external skeleton can dissolve easily in wet climates.
    (C) An external skeleton does not grow with the animal's body.
    (D) The internal skeleton prevents infections better than an external one.
    (E) Animals with external skeletons are not as successful as animals with internal skeletons.

16. All of the following are correct about animal systems EXCEPT

    (A) the hydra has a two-way digestive system
    (B) earthworms are hermaphrodites
    (C) grasshoppers use their digestive tract to remove nitrogenous waste
    (D) the nitrogenous waste from the earthworm is uric acid
    (E) the grasshopper has chitinous plates in the gizzard that act like teeth

17. Ligaments connect _____ to _____ ; tendons connect _____ to _____

      (A) bone to muscle; bone to bone
      (B) bone to bone; muscle to bone
      (C) muscle to muscle; bone to muscle
      (D) tendons to muscle; ligaments to bone
      (E) tendons to bones; muscle to muscle

Questions 18–22

      (A) Uric acid
      (B) Urea
      (C) Ammonia

18. Most toxic

19. Excreted by humans and earthworms

20. Excreted by the digestive tracts of insects and birds

21. Excreted by the Malpighian tubules

22. Excreted by nephridia

EXPLANATION OF ANSWERS

1. **(A)** A polyp looks like a long, thin cup with tentacles extending from the open end. The medusa shape of a jellyfish is a flattened and upside-down version of the same thing.

2. **(B)** In an open circulatory system, as in a grasshopper, blood leaves the artery and moves through interconnected sinuses or hemocoels, spaces surrounding the organs. This system lacks capillaries. The hydra has no circulatory system, every cell is in direct contact with the environment.

3. **(C)** In a closed circulatory system, like in an earthworm, blood travels in vessels: arteries, veins, and capillaries.

4. **(A)** The hydra has no circulatory system because every cell is in direct contact with the environment. The animal has only two cell layers, an ectoderm and an endoderm connected by a middle layer, the mesoglea, which holds the two layers together.

5. **(A)** The nervous system of the hydra is primitive. It consists of a nerve net, where impulses travel in all directions from any site.

6. **(C)** Earthworms are said to have an external respiratory surface because diffusion of carbon dioxide and oxygen gases occurs at the animal's surface across moist skin.

7. **(B)** The digestive tract of the grasshopper is responsible for removing the nitrogenous waste uric acid from the animal. Malpighian tubules serve this function.

8. **(C)** The earthworm has paired nephridia in every body segment to remove the nitrogenous waste urea.

9. **(A)** All cnidarians, including hydra and jellyfish, have stinging cells called cnidocytes.

10. **(B)** The grasshopper has an open circulatory system where blood normally leaves the artery and moves through interconnected sinuses or hemocoels, spaces surrounding the organs, before returning to a vein. This type of system lacks capillaries.

11. **(A)** The hydra has only two cell layers, an ectoderm and an endoderm connected by a middle layer, the mesoglea, which holds the two layers together. It has no circulatory system because every cell is in direct contact with the environment.

12. **(B)** Air enters the grasshopper's body through spiracles and travels through a system of tracheal tubes into the hemocoels or sinuses where diffusion occurs.

13. **(A)** The hydra is generally sessile, meaning it cannot move.

14. **(D)** Although increasing blood flow would help regulate body temperature, it is not a behavioral change.

15. **(C)** When an animal sheds its exoskeleton, it is very vulnerable to predation. There are more insects than any other animals on Earth, so the idea that its exoskeleton is a disadvantage, is without merit.

16. **(D)** The nitrogenous waste in the earthworm is urea, not uric acid. The other statements are correct.

17. **(B)** Ligaments connect bone to bone; tendons connect muscle to bone.

18. **(C)** Ammonia is the most toxic nitrogenous waste and is commonly excreted by most aquatic animals. It is very soluble in water and readily passes through membranes.

19. **(B)** Urea is less toxic than ammonia. Mammals, most adult amphibians, marine fish, and turtles excrete urea.

20. **(A)** Land snails, insects, birds, and many reptiles excrete uric acid. It is pastelike and can be excreted with very little loss of water. Conservation of water is critical for terrestrial animals.

21. **(A)** In insects, the structures of excretion are Malpighian tubules. Land snails, insects, birds, and many reptiles excrete uric acid. It is pastelike and can be excreted with very little loss of water.

22. **(B)** The earthworm has paired nephridia in every body segment, and these excrete urea.

Human Physiology

• Digestion   • Nervous system • Gas exchange   • Eye and ear • Circulation   • Excretion • Endocrine system   • Muscles

This chapter includes a review of the following topics about human physiology—digestion, gas exchange, circulation, endocrine system, nervous system, excretion, and muscles,. It also includes 45 review questions. To maximize the number of review questions, two remaining topics in human physiology are included in the next chapter.

DIGESTION

The human digestive system has two important functions: breaking down large food molecules into smaller, usable molecules and absorbing these smaller molecules. Fats get broken down into glycerol and fatty acids, starch into monosaccharides, nucleic acids into nucleotides, and proteins into amino acids. Vitamins and minerals are small enough to be absorbed without being digested.

The digestive tract is about 30 feet long and made of smooth (involuntary) muscle that pushes the food along the digestive tract by a process called peristalsis. The muscles of the digestive tract are controlled by the autonomic nervous system. Figure 15.1 is a chart detailing the structures and function of the human digestive system.

Mouth

- Mechanical and chemical digestion begins here.
- The enzyme salivary amylase in saliva begins starch digestion.
- The tongue and differently shaped teeth work together to break down food mechanically.
- The type of teeth an animal has is a reflection of its dietary habits. Humans are omnivores and have three different types of teeth: incisors for cutting, canines for tearing, and molars for grinding.

Esophagus

- No digestion occurs here.
- After swallowing, food is directed into the esophagus and away from the windpipe by the **epiglottis**, a flap of cartilage in the back of the **pharynx** (throat).
- The esophagus transports food from the throat to the stomach.

> **REMEMBER**
>
> Starch digestion begins in the mouth.

217

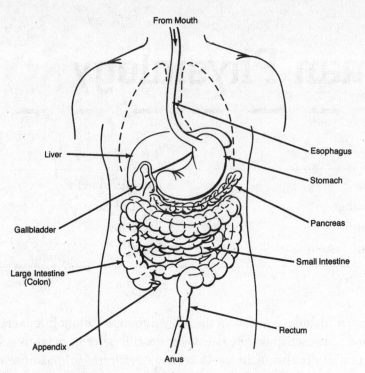

Figure 15.1 Human digestive system

Stomach

- Both mechanical and chemical digestion occur here.
- Protein digestion begins in the stomach.
- The stomach's thick, muscular wall churns food mechanically and secretes gastric juice, which contains hydrochloric acid and enzymes that digest proteins.
- Hydrochloric acid begins the breakdown of muscle (meat) and activates the inactive enzyme pepsinogen to become pepsin, which digests protein.
- The enzyme rennin aids in the digestion of the protein in milk.
- The pH in the stomach is acidic, ranging from 2 to 3.
- The cardiac sphincter at the top of the stomach keeps acidified food in the stomach from backing up into the esophagus and burning it. The pyloric sphincter at the bottom of the stomach keeps the food in the stomach long enough to be digested.
- Excessive acid can cause an ulcer to form in the esophagus, the stomach, or the duodenum (the upper intestine). We now know that a common cause of ulcers is a particular bacterium, *Heliobacter pylori*, which can be effectively treated with antibiotics.

REMEMBER

Protein digestion begins in the stomach.

Small Intestine

- All digestion is completed and nutrients are absorbed here.
- The pH in the small intestine is 8.
- It is 6 meters long.
- All digestion is completed in the duodenum, the first 10 inches of small intestine.
- The intestinal enzymes are amylases, proteases, lipases, and nucleases.
- Pancreatic amylases, which digest starch, are secreted into the small intestine.
- **Peptidases**, such as trypsin and chymotrypsin, continue to break down proteins.
- Nucleases hydrolyze nucleic acids into nucleotides.
- Lipases break down fats.
- Millions of fingerlike projections called villi line the small intestine and absorb all nutrients that were previously released from digested food.
- Each **villus** contains capillaries, which absorb amino acids, vitamins, and monosaccharides directly into the bloodstream, and a **lacteal**, which absorbs fatty acids and glycerol into the lymphatic system.
- Villi have microscopic appendages called microvilli that further enhance the rate of absorption. Figure 15.2 is a drawing of a villus.

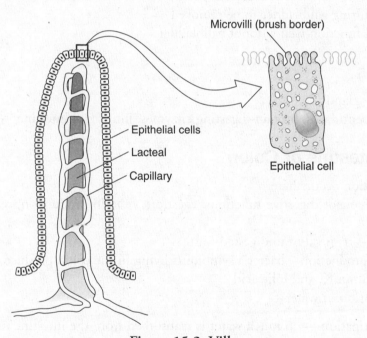

Figure 15.2 Villus

Liver

- Accessory gland
- Produces **bile** that emulsifies fats.
- Bile = pH 11; neutralizes chyme (acidified food from stomach) entering small intestine
- Sends bile to the gallbladder until its release into the small intestine
- Has other functions besides digestion

  1. Breaks down and recycles red blood cells
  2. Detoxifies blood—removes alcohol and drugs
  3. Produces cholesterol necessary for structure of cell membranes
  4. Produces the nitrogenous waste urea from protein metabolism

Gallbladder

- Accessory gland
- Stores bile that is produced in liver
- Bile emulsifies fats in small intestine
- Gallstones (abnormal cholesterol deposits) block normal function and cause great pain, requiring gallbladder to be removed
- Body can function well without gallbladder

Pancreas

- Accessory gland
- Secretes **peptidases**, protein-digesting enzymes, into small intestine

Large Intestine or Colon

- No digestion occurs here
- Has three major digestive functions: egestion, vitamin production, reabsorption of water
- **Egestion**—removal of undigested waste
- Vitamin production—bacteria symbionts living in the colon produce the B vitamins, vitamin K, and folic acid
- **Reabsorption** of water—

  1. Constipation—too much water is reabsorbed from the intestine into body
  2. Diarrhea—an inadequate amount of water is absorbed back into body

Rectum

- Egestion—removal of undigested waste
- Last 7 to 8 inches of the gastrointestinal tract stores feces until their release through the anus

Hormones That Regulate the Digestive System

Digestive hormones are released as needed, as we see or smell food, or as food moves along the gut. Table 15.1 is a summary of the hormones involved in regulating digestion.

TABLE 15.1

Digestive Hormones		
Hormone	**Site of Production**	**Effect**
Gastrin	Stomach wall	Stimulates sustained secretion of gastric juice
Secretin	Duodenum wall	Stimulates pancreas to release bicarbonate to neutralize acid in duodenum
Cholecystokinin (CCK)	Duodenum wall	Stimulates pancreas to release pancreatic enzymes and gallbladder to release bile into small intestine

GAS EXCHANGE

In humans, air enters the nasal cavity and is moistened, warmed, and filtered. From there, air passes through the larynx and down the trachea and bronchi into the tiniest bronchioles, which end in microscopically tiny air sacs called **alveoli**. Here is where diffusion of respiratory gases occurs.

Humans have an internal respiratory surface because respiratory gases are exchanged deep inside the body. The rib cage expands and forces the diaphragm to contract and move downward, thus expanding the chest cavity and decreasing the internal pressure. Air is drawn into the lungs by negative pressure because the internal pressure inside the chest cavity is lower than the air pressure surrounding the body.

The medulla in the brain sets the breathing rhythm by monitoring carbon dioxide levels in the blood and by sensing changes in the pH of the blood. A blood pH lower than 7.4 triggers autonomic nerves from the medulla to increase the breathing rate to rid the body of more carbon dioxide. The concentration of oxygen in the blood usually has little effect on the breathing control centers.

Hemoglobin

Oxygen is carried in the human blood by the respiratory pigment hemoglobin, which combines loosely with oxygen molecules to form the molecule oxyhemoglobin. To function in the transport of oxygen, hemoglobin must be able to bind with oxygen in the lungs and unload it at the body cells. That means that the more tightly the hemoglobin binds to oxygen in the lungs, the more difficult it is to unload the oxygen at the cells.

Transport of Carbon Dioxide

Carbon dioxide, the by-product of cell respiration, is released from every cell and dissolves in the blood to form carbonic acid. Therefore, the higher the carbon dioxide concentration in the blood, the lower the pH.

Very little carbon dioxide is transported by hemoglobin. Most carbon dioxide is carried in the plasma as part of the reversible blood-buffering carbonic acid-bicarbonate ion system, which maintains the blood at a constant pH of 7.4.

CIRCULATION

Human circulation consists of a closed circulatory system with arteries, veins, and capillaries. Table 15.2 shows the characteristics of each structure.

TABLE 15.2

Structure and Function of Blood Vessels

Vessel	Function	Structure
Artery	Carries blood away from the heart under enormous pressure	Walls made of thick layer of elastic, smooth muscle. Can withstand high pressure and can contract and expand as needed.
Vein	Carries blood back to the heart under very little pressure	Walls do not contain thick layer of muscle. Has valves to help prevent backflow. Located within skeletal muscle, which propels blood upward and back to heart as the body moves and muscles contract.
Capillary	Allows for diffusion of nutrients and wastes between cells and blood	Walls are one-cell thick and so small that blood cells travel only single file. Blood travels slowly here to allow time for diffusion of nutrients and wastes.

Components of Blood

Blood consists of several different cell types suspended in a liquid matrix called **plasma.** The average human body contains 4 to 6 liters of blood. Table 15.3 is a chart that gives basic information about blood.

TABLE 15.3

Components of Blood

Component	Scientific Name	Properties
Plasma	None	Liquid portion of the blood.
		Contains clotting factors, hormones, antibodies, dissolved gases, nutrients, and wastes.
		90% water.
Red blood cells	**Erythrocytes**	Carry hemoglobin and oxygen.
		Do not have a nucleus.
		Live about 120 days.
		Formed in the bone marrow and recycled in the liver.
White blood cells	**Leukocytes**	Fight infection.
		Formed in the bone marrow.
		Die fighting infection and are one component of pus.
		Most common types of leukocytes: neutrophils and lymphocytes.
Platelets	**Thrombocytes**	Clot blood.
		Cell fragments that are formed in the bone marrow from megakaryocytes.

The Mechanism of Blood Clotting

Blood clotting is a complex mechanism that begins with the release of clotting factors from platelets and damaged tissue. It involves a complex set of reactions. Anticlotting factors constantly circulate in the plasma to prevent the formation of a clot or thrombus, which can cause serious damage in the absence of injury. **Serum** is plasma minus clotting factors. Calcium is necessary for normal blood clotting.

Here is the pathway of normal clot formation:

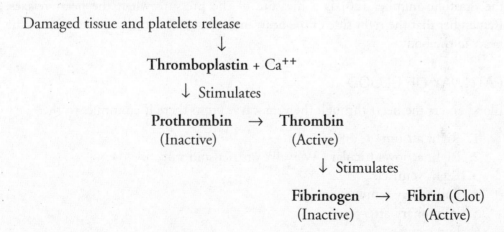

Damaged tissue and platelets release
↓
Thromboplastin + Ca^{++}
↓ Stimulates
Prothrombin → **Thrombin**
(Inactive)   (Active)
↓ Stimulates
Fibrinogen → **Fibrin** (Clot)
(Inactive)   (Active)

The Heart

The heart is located beneath the sternum and is about the size of your clenched fist. Figure 15.3 is a diagram of the heart. It beats about 70 beats per minute and pumps about 5 liters of blood—the total volume of blood in the body—each minute. Two atria (atria, plural; atrium, singular) receive blood from the cells of the body, and two ventricles pump blood out of the heart.

The heart itself has its own pacemaker, the **sinoatrial (SA) node**, which sets the timing of the contractions of the heart. Electrical impulses travel through the cardiac and body tissues to the skin, where they can be detected by an electrocardiogram (EKG). The heart's pacemaker is influenced by a variety of factors: the nervous system, hormones such as adrenaline, and body temperature.

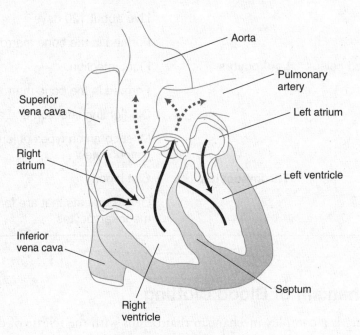

Figure 15.3 The heart

Blood pressure is lowest in the veins and highest in the arteries when the ventricles contract. Blood pressure for all normal resting adults is 120/80. The **systolic** number (120) is a measurement of the pressure when the ventricles contract, while the **diastolic** number (80) is a measure of the pressure when the heart relaxes. Remember that the right side of the heart in the figure is actually the left side of the heart in the body.

PATHWAY OF BLOOD

Blood enters the heart through the vena cava. From there it continues to the:

1. Right atrium
2. Right atrioventricular (AV) valve or tricuspid valve
3. Right ventricle
4. Pulmonary semilunar valve
5. Pulmonary artery
6. Lungs

STUDY TIP

Make sure you know the pathway of the blood throughout the body.

224

SAT Biology

7. Pulmonary vein
8. Left atrium
9. Left atrioventricular (AV) valve or bicuspid valve
10. Left ventricle
11. Aortic semilunar valve
12. Aorta
13. To all the cells in the body
14. Returns to the heart through the vena cava

Blood circulates through the coronary circulation (heart), renal circulation (kidneys), and the hepatic circulation (liver). The pulmonary circulation includes the pulmonary artery, lungs, and the pulmonary vein.

One thing to remember is that the pulmonary artery is the only artery that carries deoxygenated blood and the pulmonary vein is the only vein that carries oxygenated blood. Figure 15.4 is a drawing of the human circulatory system.

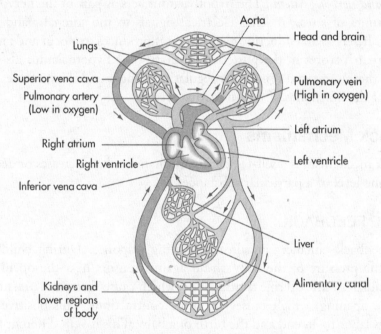

Figure 15.4 Human circulation

ENDOCRINE SYSTEM

Animals have two major regulatory systems that release chemicals: the endocrine system and the nervous system. Even though the two systems are separate, they work together to regulate the body, to maintain **homeostasis**. The endocrine system secretes hormones, while the nervous system secretes neurotransmitters. In one case, epinephrine functions both as the fight-or-flight hormone secreted by the adrenal gland and as a neurotransmitter that sends a message from one neuron to another.

Hormones are produced in ductless (endocrine) glands and move through the blood to a specific target cell, tissue, or organ that can be far from the original endocrine gland. They can produce either an immediate short-lived response. An example of a short-lived response is the way adrenaline (epinephrine) causes the flight-or-fight response, or a dramatic long-term development of an entire organism. An example of a long-term response is the way ecdysone controls metamorphosis in insects.

Tropic hormones are hormones that stimulate other glands to release hormones and can have a far-reaching effect. For example, the anterior pituitary in the brain releases thyroid-stimulating hormone (TSH), which stimulates the thyroid in the neck region to release thyroxin. Other types of chemical messengers reach their target by special means. Pheromones in the urine of a dog carry a message between different individuals of the same species. In vertebrates, nitric oxide (NO), a gas, is produced by one cell and diffuses to and affects only neighboring cells before it is broken down.

The Hypothalamus

The **hypothalamus** plays a special role in the body; it is the *bridge between the endocrine and nervous systems*. The hypothalamus acts as part of the nervous system when, in times of stress, it sends electrical signals to the adrenal gland to release adrenaline. It acts as an endocrine gland when it produces oxytocin and antidiuretic hormone that it stores in the posterior pituitary. The hypothalamus also contains the body's thermostat and centers for regulating hunger and thirst. Table 15.4 is an overview of the hormones of the endocrine system.

Feedback Mechanisms

A feedback mechanism is a self-regulating mechanism that increases or decreases an action or the level of a particular substance.

POSITIVE FEEDBACK

Positive feedback enhances an already existing response. During childbirth, for example, the pressure of the baby's head against sensors near the opening of the uterus stimulates more uterine contractions, which causes increased pressure against the uterine opening, which causes yet more contractions. This positive feedback loop brings labor to an end and the birth of a baby. This is very different from negative feedback.

NEGATIVE FEEDBACK

Negative feedback is a common mechanism in the endocrine system (and elsewhere) that maintains homeostasis. A good example is how the body maintains proper levels of thyroxin. When the level of thyroxin in the blood is too low, the hypothalamus stimulates the anterior pituitary to release a hormone, thyroid-stimulating hormone (TSH), which stimulates the thyroid to release more thyroxin. When the level of thyroxin is adequate, the hypothalamus stops stimulating the pituitary.

TABLE 15.4

Endocrine Hormones

Gland	Hormone	Effect
Anterior pituitary	• Growth hormone (GH)	Stimulates growth of bones
	• Luteinizing hormone (LH)	Stimulates ovaries and testes
	• Thyroid-stimulating hormone (TSH)	Stimulates thyroid gland
	• Adrenocorticotropic (ACTH) hormone	Stimulates adrenal cortex to secrete glucocorticoids
	• Follicle-stimulating hormone (FSH)	Stimulates gonads to produce sperm and ova
Posterior pituitary	• Oxytocin	Stimulates contractions of uterus and mammary glands
	• Antidiuretic hormone (ADH)	Promotes retention of water by kidneys
Thyroid	• Thyroxin	Controls metabolic rate
	• Calcitonin	Lowers blood calcium levels
Parathyroid	Parathormone	Raises blood calcium levels
Adrenal cortex	Glucocorticoids	Raises blood sugar levels
Adrenal medulla	• Epinephrine (adrenaline)	Raises blood sugar level by increasing rate of glycogen breakdown by liver
	• Nonepinephrine (noradrenaline)	
Pancreas—islets of Langerhans	• Insulin	Lowers blood glucose levels
	• Glucagon	Raises blood glucose levels
Thymus	Thymosin	Stimulates T lymphocytes as part of the immune response
Pineal	Melatonin	Involved in biorhythms
Ovaries	• Estrogen	Stimulates uterine lining, promotes development and maintenance of primary and secondary characteristics of female
	• Progesterone	Promotes uterine lining growth
Testes	Androgens	Support sperm production and promote secondary sex characteristics

How Hormones Trigger a Response in Target Cells

There are two types of hormones, and they stimulate target cells in different ways. These are illustrated in Figure 15.5.

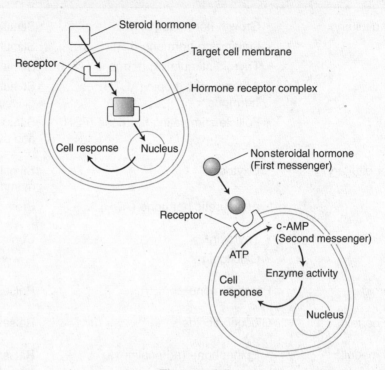

Figure 15.5

Lipids or steroid hormones diffuse directly through the plasma membrane and bind to a receptor inside the cell that triggers the cell's response.

Protein or polypeptide hormones (nonsteroidal) cannot dissolve in the plasma membrane so they bind to a receptor on the surface of the cell. Once the hormone (the first messenger) binds to a receptor on the surface of the cell, it triggers a secondary messenger such as c-AMP, which converts the extracellular chemical signal to a specific response inside the cell.

NERVOUS SYSTEM

The vertebrate nervous system consists of central and peripheral components.

- The central nervous system (CNS) consists of the brain and spinal cord.
- The peripheral nervous system (PNS) consists of all nerves outside the CNS.

The peripheral nervous system is then further divided and subdivided into various systems. The following is an overview of the peripheral nervous system.

Outline of the Peripheral Nervous System

Sensory—Conveys information from sensory receptors or nerve endings
Motor—Stimulates voluntary and involuntary muscles

Consists of two systems:

Somatic System   Controls the voluntary muscles
Autonomic System   Controls involuntary muscles

Sympathetic

- Fight-or-flight response
- Increases heart and breathing rate
- Liver converts glycogen to glucose
- Bronchi of lungs dilate and increase gas exchange
- Adrenaline raises blood glucose levels

Parasympathetic

- Opposes the sympathetic system
- Calms the body
- Decreases heart/breathing rate
- Enhances digestion

The Neuron

The neuron is the basic functional unit of the nervous system and is illustrated in Figure 15.6. It consists of a cell body which contains the nucleus and other organelles, and two types of cytoplasmic extensions, called dendrites and axons.

- Dendrites are sensory. They receive incoming messages from other cells and carry the electrical signal to the cell body. A neuron can have hundreds of dendrites.
- Axons transmit an impulse from the cell body outward to another cell. A neuron has only one axon, which can be several feet long in large mammals. Most axons are wrapped in a myelin sheet that protects the axon and speeds the impulse.

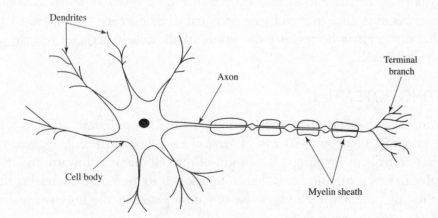

Figure 15.6 The neuron

The Reflex Arc

The simplest nerve response is a reflex arc. It is inborn, automatic, and protective. An example is the knee-jerk reflex, which consists of only a sensory and a motor neuron. You have experienced this during a physical exam. The doctor taps your knee with a hammer, and your foot kicks up involuntarily. The impulse moves from the sensory neuron in your knee to the motor neuron that directs the thigh muscle to contract. The spinal cord is not involved in this type of reflex.

A more complex reflex arc consists of three neurons: a sensory, a motor, and an interneuron or association neuron. This is illustrated in Figure 15.7. A sensory neuron transmits an impulse to the interneuron in the spinal cord which sends one impulse to the brain for processing and also one to the motor neuron to effect change immediately (at the muscle). This is the type of response that quickly jerks your hand away from a hot iron before your brain has figured out what occurred.

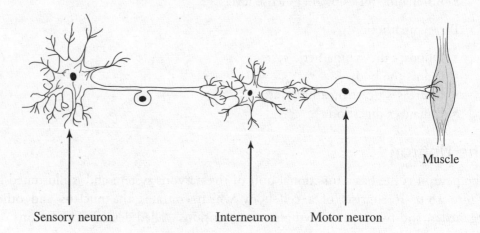

Sensory neuron   Interneuron   Motor neuron

Muscle

Figure 15.7 The reflex arc

How a Neuron Functions

All living cells exhibit a membrane potential, a difference in electrical charge between the cytoplasm (negative ions) and extracellular fluid (positive ions). Physiologists measure this difference in membrane potential to be between –50mV to –100mV. The negative sign indicates that the inside of the cell is negative relative to the outside of the cell.

RESTING POTENTIAL

A neuron at rest or unstimulated (resting potential) is **polarized** and has a membrane potential of about –70 mV. The sodium-potassium pump maintains this polarization by actively pumping ions out of the cell that leak inward. In order for the nerve to fire, a stimulus must be strong enough to overcome the resting threshold or resting potential. *The larger the membrane potential, the stronger the stimulus must be to cause the nerve to fire.*

ACTION POTENTIAL

An action potential, or impulse, can only be generated in the axon of a neuron. When an axon is stimulated sufficiently to overcome the threshold, the permeability of a region of the membrane suddenly changes. Sodium channels open and sodium floods into the cell, down the concentration gradient. In response, potassium channels open and potassium floods out of the cell. This rapid movement of ions or wave of depolarization reverses the polarity of the membrane and is called an action potential. The action potential is localized and lasts a very short time.

The sodium-potassium pump restores the membrane to its original polarized condition by pumping sodium and potassium ions back to their original positions. This period of repolarization, which lasts a few milliseconds, is called the refractory period, during which the neuron cannot respond to another stimulus. The refractory period ensures that an impulse moves along an axon in one direction only since the impulse can move only to a region where the membrane is polarized. Figure 15.8 shows the axon of a neuron as an impulse passes from left to right, depolarizing the membrane in front of it.

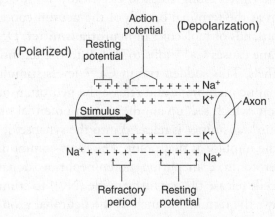

Figure 15.8 An impulse passing along an axon

The action potential is like a row of dominoes falling in order after the first one is knocked over. The first action potential generates a second action potential that generates a third, and so on. The impulse moves along the axon, propagating itself *without losing any strength*. If the axon is myelinated, the impulse travels faster.

The action potential is an all-or-none event. Either the stimulus is strong enough to cause an action potential, or it is not. The body distinguishes between a strong stimulus and a weak one by the frequency of action potentials. A strong stimulus sets up more action potentials than a weak one does.

Figure 15.9 traces the events of a membrane undergoing sufficient stimulation to undergo an action potential.

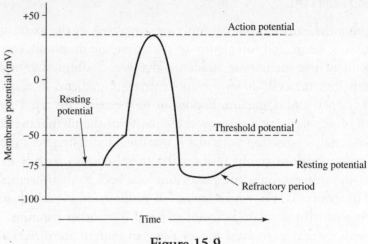

Figure 15.9

THE SYNAPSE

Although an impulse travels along an axon electrically, it crosses a synapse chemically. The cytoplasm at the terminal branch of the neuron contains many **vesicles,** each containing thousands of molecules of **neurotransmitter**. Depolarization of the presynaptic membrane causes Ca^{++} ions to rush into the terminal branch through calcium-gated channels. This sudden rise in Ca^{++} levels stimulates the vesicles to fuse with the presynaptic membrane and release the neurotransmitter by exocytosis into the synaptic cleft, which sets up another action potential on the adjacent cell. Shortly after the neurotransmitter is released into the synapse, it is destroyed by an enzyme that stops the impulse at that point. The most common neurotransmitters are acetylcholine, serotonin, epinephrine, norepinephrine, dopamine, and GABA. In addition, many cells release the gas nitric oxide (NO) to stimulate other cells.

Figure 15.10 shows the terminal branch of the neuronal axon and the synapse.

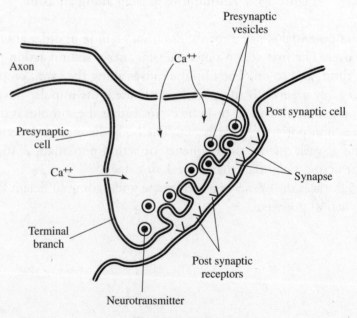

Figure 15.10

The Eye and the Ear

Occasionally, a diagram of the eye or the ear finds its way onto an SAT II Subject Test. You should know the basic parts of each. This section lists the structures and functions of each that you should know, and Figure 15.11 is an illustration.

EYE

- **Cones**—photoreceptors in the retina that distinguish different colors
- **Cornea**—tough, clear covering that protects the eye and allows light to pass through
- **Humor**—fluids that maintain the shape of the eyeball
- **Iris**—colored part of the eye that controls how much light enters the eye
- **Lens**—focuses light onto the retina
- **Pupil**—small opening in the middle of the iris
- **Retina**—converts light into nerve impulses that are carried to the brain
- **Rods**—photoreceptors in the retina that are extremely sensitive, but do not distinguish different colors

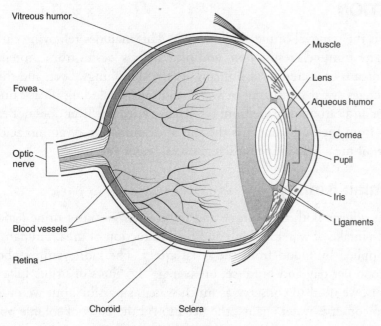

Figure 15.11 The human eye

EAR

- **Auditory canal**—ear canal, where sound enters
- **Cochlea**—fluid-filled part of inner ear, sends nerve impulses to brain
- **Ear bones**—hammer, anvil, and stirrup; transmit vibrations from eardrum to oval window
- **Eustachian tube**—equalizes pressure between environment and inner ear
- **Oval window**—sends waves of pressure to the cochlea
- **Semicircular canals**—fluid filled, helps you maintain your balance
- **Tympanum** - ear drum, vibrates as sound waves hit it

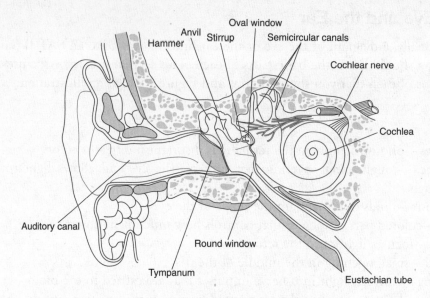

Figure 15.12 The human ear

STUDY TIP
The organs of excretion are: • skin • lungs • kidney • liver

EXCRETION

Excretion is the removal of metabolic wastes. This includes removing carbon dioxide and water from cell respiration, and nitrogenous wastes from protein metabolism. The organs of excretion in humans are the skin, lungs, liver, and the kidneys.

Skin excretes sweat consisting of water and salts, including urea. Lungs excrete water vapor and carbon dioxide from the Krebs cycle. The liver does not excrete any substances from the body, but it is the site of deamination of amino acids and the production of urea. The kidneys excrete excess water and urea.

The Human Kidney

The kidney adjusts both the volume and the concentration of urine depending on the animal's intake of water and salt and the production of urea. Humans have two kidneys supplied by blood from the renal artery. The kidneys filter about 1,500 liters of blood per day and produce, on average, 1.5 liters of urine. Like all terrestrial animals, we need to conserve as much water as possible, but we must balance the need to conserve water against the need to rid the body of soluble poisons.

THE NEPHRON

The nephron, as shown in Figure 15.13, is the basic functional unit of the kidney. It consists of a cluster of capillaries, known as the glomerulus—which sits inside a cuplike structure called Bowman's capsule, a long narrow tube called the tubule, and the loop of Henle. Each human kidney contains about 1 million nephrons. The nephron carries out its job in four steps: filtration, secretion, reabsorption, and excretion.

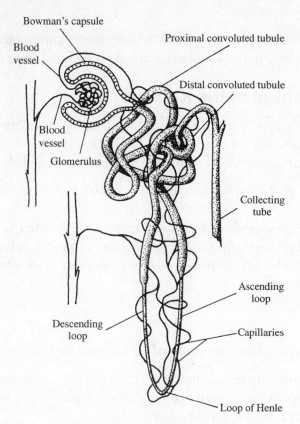

Figure 15.13 The nephron

Filtration

Filtration occurs by *diffusion*. It is *passive* and *nonselective*. The filtrate contains everything small enough to diffuse out of the glomerulus and into Bowman's capsule, including glucose, salts, vitamins, wastes such as urea, and other small molecules. From Bowman's capsule, the filtrate travels into the loop of Henle and then the collecting duct or tubule. From the collecting tubule, the filtrate trickles into the ureter and the urinary bladder for temporary storage and then to the **urethra** and out of the body.

Secretion

Secretion is the *active, selective* uptake of molecules that did not get filtered into Bowman's capsule. This occurs in the tubules of the nephron.

Reabsorption

Reabsorption is the process by which most of the water and solutes (glucose, amino acids, and vitamins) that initially entered the tubule during filtration are transported back into the capillaries and thus, back to the body. This process occurs in the tubule, the loop of Henle, and the collecting tubule. The longer the loop of Henle, the greater is the reabsorption of water.

Excretion

Excretion is the removal of metabolic wastes, for example, nitrogenous wastes. Everything that passes into the collecting tubule is excreted from the body.

Hormone Control of the Kidneys

The kidney is able to respond quickly to the changing requirements of the body because it is it under the control of the endocrine system.

Antidiuretic hormone, ADH, regulates the concentration of the blood.

* It is produced in the hypothalamus and is stored and released by the posterior pituitary.
* It increases the permeability of the collecting tubule to water so that more water can be reabsorbed back into the body and urine volume is reduced.
* Drinking alcohol blocks the release of ADH, resulting in increased urination and even dehydration.

MUSCLES

There are three types of muscle: smooth, cardiac, and skeletal.

* Smooth or involuntary muscle makes up the walls of blood vessels and the digestive tract. It does not have a striated appearance, hence the name. It is under the control of the autonomic nervous system.
* Cardiac muscle is found in the heart and is not striated. It generates its own action potential. Individual heart cells will beat on their own, even when removed from the body.
* Skeletal or voluntary (striated) muscles are very large and multinucleate. They work in pairs, one muscle contracts while the other relaxes. The biceps and triceps in your arm are one such pair.

The Sliding Filament Theory

Within the cytoplasm of each skeletal muscle cell are thousands of fibers called myofibrils that run parallel to the length of the cell. Myofibrils consist of thick and thin filaments. Each thin filament consists of **actin** proteins; each thick filament is composed of **myosin** proteins. Muscles contract as thick and thin filaments slide over each other.

MULTIPLE-CHOICE QUESTIONS

1. All of the following are functions of the liver EXCEPT

   (A) cholesterol production
   (B) recycles red blood cells
   (C) site of bile production
   (D) site of deamination of amino acids
   (E) excretion of urea

2. Absorption of nutrients occurs in the

   (A) stomach
   (B) beginning of the small intestine
   (C) latter part of the small intestine
   (D) colon
   (E) rectum

3. Which pair is correctly matched?

   (A) Pyloric sphincter—small intestine
   (B) Colon—villi
   (C) Pancreas—bile
   (D) *H. pylori*—ulcers
   (E) Amylase—digests proteins

Questions 4–10

Match the description to the parts of the human digestive system below.

   (A) Small intestine
   (B) Stomach
   (C) Esophagus
   (D) Colon
   (E) Mouth

4. Digestion of starch begins

5. Digestion of starch is completed

6. Digestion of proteins begins

7. Water absorption

8. Vitamin production

9. Organ of egestion

10. Contains villi

11. Gastric enzymes work best in an environment with a pH of

    (A) 3
    (B) 6
    (C) 7
    (D) 8
    (E) 11

12. Intestinal enzymes work best in an environment with a pH of

    (A) 2
    (B) 6
    (C) 7
    (D) 8
    (E) 11

13. The lacteal is found in the _____ and is involved with

    _____ .

    (A) stomach; release of hormones
    (B) duodenum; hydrolysis of lipids
    (C) villi; the absorption of fatty acids
    (D) colon; the reabsorption of water
    (E) liver; production of hydrolytic enzymes

14. _____ is a hormone released by the duodenum
    wall that stimulates the _____ to release bicarbonate to
    neutralize acid in the duodenum.

    (A) Gastrin; stomach wall
    (B) Gastrin; pancreas
    (C) Secretin; stomach wall
    (D) Secretin; pancreas
    (E) CCK; small intestine

15. The digestive tract consists of _____ and is under the
    control of the _____ nervous system.

    (A) smooth muscle; autonomic
    (B) smooth muscle; somatic
    (C) skeletal muscle; autonomic
    (D) skeletal muscle; somatic
    (E) striated muscle; autonomic

16. All of the following are correct about gas exchange in humans EXCEPT

    (A) hemoglobin carries oxygen and carbon dioxide in fairly equal amounts
    (B) breathing rate is regulated by the medulla of the brain
    (C) the purpose of the nose is to filter, warm, and moisten air
    (D) gas exchange occurs in alveoli
    (E) when carbon dioxide builds up in the blood, the pH lowers

17. All of the following are correct about human red blood cells EXCEPT

    (A) they live 120 days
    (B) they are called leukocytes
    (C) they do not have a nucleus
    (D) they are formed in the bone marrow
    (E) they carry oxyhemoglobin

18. All of the following are correct about human circulation EXCEPT

    (A) arteries have thick muscular walls
    (B) the pacemaker of the heart is the SA node
    (C) leukocytes fight disease
    (D) hemoglobin is allosteric and changes its conformation in response to changes in the environment
    (E) the only artery carrying deoxygenated blood is the aorta

19. All of the following are related to blood clotting EXCEPT

    (A) fibrinogen
    (B) erythrocytes
    (C) thromboplastin
    (D) calcium ions
    (E) megakaryocytes

20. All of the following are correct about the normal direction of blood flow EXCEPT

    (A) vena cava $\rightarrow$ right ventricle
    (B) right atrium $\rightarrow$ tricuspid valve
    (C) aortic semilunar valve $\rightarrow$ aorta
    (D) lungs $\rightarrow$ pulmonary vein
    (E) pulmonary vein $\rightarrow$ left atrium

21. Which of the following correctly follows the flow of blood in the human body?

    (A) Right ventricle—aorta—body—left atrium—left ventricle—pulmonary artery

    (B) Right ventricle—pulmonary artery—lungs—pulmonary vein—left atrium

    (C) Left atrium—left ventricle—pulmonary artery—pulmonary vein—right atrium

    (D) Left atrium—left ventricle—pulmonary artery—lungs—pulmonary vein—right atrium

    (E) Left ventricle—left atrium—aorta—lungs—pulmonary vein

22. All of the following are correct about human circulation EXCEPT

    (A) blood in capillaries travels slowly to allow for diffusion of nutrients and wastes

    (B) normal blood pressure is higher in males than in females

    (C) the average heart rate and pulse is about 70 beats per minute

    (D) your heart rate normally changes in response to physical activity

    (E) arteries have thicker walls than veins

23. The largest amount of the carbon dioxide is carried to the lungs by

    (A) leukocytes

    (B) thrombocytes

    (C) erythrocytes

    (D) hemoglobin in red blood cells

    (E) the plasma as the bicarbonate ion

24. All of the following are true about blood EXCEPT

    (A) white blood cells are formed in the bone marrow

    (B) platelets are not cells but are cell fragments

    (C) arteries contain valves to help in the pumping of blood

    (D) the liquid portion of the blood is called plasma

    (E) red blood cells live about 120 days

Questions 25–29

Match the hormone to the correct gland from which it comes.

      (A) Thyroxin
      (B) Oxytocin
      (C) Insulin
      (D) Thyroid-stimulating hormone
      (E) Adrenaline

25. Pancreas

26. Anterior pituitary

27. Posterior pituitary

28. Adrenal medulla

29. Thyroid

Questions 30–34

Match the hormone and its function. Use each letter once only.

      (A) Raises blood sugar
      (B) Lowers blood sugar
      (C) Stimulates the ovaries
      (D) Stimulates the uterine lining
      (E) Stimulates growth of long bones

30. Progesterone

31. HGH

32. Glucagon

33. FSH

34. Insulin

35. All of the following are correct about the endocrine system EXCEPT

    (A) endocrine glands are ductless
    (B) parathormone is the fight-or-flight hormone
    (C) the islet cells in the pancreas produce both glucagon and insulin
    (D) ecdysone controls metamorphosis in some insects
    (E) nonpolar hormones must bind with a receptor and trigger a secondary messenger

36. All of the following hormones are produced by the anterior pituitary EXCEPT

    (A) glucagon
    (B) follicle-stimulating hormone
    (C) human growth hormone
    (D) thyroid-stimulating hormone
    (E) adrenocorticotropic hormone

37. The main target of antidiuretic hormone is the

    (A) liver
    (B) kidney
    (C) heart
    (D) thyroid
    (E) spleen

38. Tropic hormones

    (A) are secreted by the thyroid
    (B) are secreted by the liver
    (C) are released by the hypothalamus
    (D) are released only in females
    (E) are hormones that stimulate other glands

39. Which of the following is correct about the kidney?

    (A) Filtration occurs in the Bowman's capsule and is nonselective.
    (B) Filtration is selective and occurs by active transport.
    (C) Filtration is selective and occurs by passive transport.
    (D) Reabsorption is selective and occurs by passive transport only.
    (E) Reabsorption occurs in the glomerulus.

40. In humans, urea is

    (A) produced in the kidney and excreted by the skin
    (B) produced in the liver and excreted mainly by the kidneys
    (C) produced in the kidney from the metabolism of fats
    (D) produced in the lungs as a by-product of exhaling
    (E) accumulates in our bodies as we age

41. Which is true about the human circulatory system?

    (A) The pH varies throughout the day and night.
    (B) Erythrocytes clot blood.
    (C) Leukocytes fight germs.
    (D) Blood type O is dominant.
    (E) Humans have an open circulatory system.

42. ADH most directly affects the

    (A) glomerulus
    (B) Bowman's capsule
    (C) thymus
    (D) lacteal
    (E) collecting tubule

43. The following are all correct about the nervous system EXCEPT

    (A) dendrites are sensory
    (B) axons carry an impulse away from the cell body
    (C) the parasympathetic nervous system is associated with
        fight-or-flight
    (D) the central nervous system consists of the brain and spinal cord
    (E) the interneuron is located in the spinal cord

44. Which would occur if the parasympathetic system were stimulated?

    (A) Increase in blood sugar
    (B) Increase in digestion
    (C) Increase in breathing rate
    (D) Increase in adrenaline
    (E) Increase in epinephrine

45. The walls of arteries consist of

    (A) striated muscle and are under voluntary control
    (B) striated muscle and are not under voluntary control
    (C) smooth muscle and are controlled by the somatic nervous system
    (D) smooth muscle and are controlled by the autonomic nervous system
    (E) a mixture of striated and smooth muscle under control of the
        autonomic nervous system

EXPLANATION OF ANSWERS

1. **(E)** The kidneys excrete almost all of the urea we produce. The skin excretes a small amount. The liver is an accessory organ and does not excrete anything. It is, however, the organ where urea is produced by deamination of amino acids.

2. **(C)** All digestion is completed in the duodenum, the first 10 inches of the small intestine. After that the nutrients are small enough to be absorbed into the bloodstream. This occurs in the latter part of the small intestine through villi and microvilli.

3. **(D)** The bacterium *Heliobacter pylori* has been found to be the major cause of most ulcers. The pyloric sphincter is at the top of the stomach. The colon is another name for the large intestine, and villi are located in the small intestine. The liver produces bile. The pancreas contains the islets of Langerhans, which produce insulin and glucagon. Amylase is an enzyme that digests starch.

4. **(E)** Starch digestion begins in the mouth and is completed in the small intestine.

5. **(A)** Starch digestion begins in the mouth and is completed in the small intestine.

6. **(B)** Digestion of proteins begins in the stomach with gastric juice consisting of hydrochloric acid and pepsinogen. Protein digestion is completed in the small intestine.

7. **(D)** The colon or large intestine has several functions: egestion of undigested food, reabsorption of excess water, and vitamin production. Symbiotic bacteria that live in the colon produce vitamins we require.

8. **(D)** Symbiotic bacteria that live in the colon produce vitamins we require: B vitamins, vitamin K, and folic acid.

9. **(D)** The colon or large intestine has several functions: egestion of undigested food, reabsorption of excess water, and vitamin production.

10. **(A)** Millions of fingerlike projections called villi line the small intestine and absorb all nutrients that were previously released from digested food. Each villus contains capillaries that absorb amino acids, vitamins, and monosaccharides directly into the bloodstream and a lacteal, which absorbs fatty acids and glycerol into the lymphatic system. Villi have microscopic appendages called microvilli that further enhance the rate of absorption.

11. **(A)** Different enzymes have different requirements. Gastric (stomach) enzymes work in the stomach where the pH is very acidic, a pH of 3. Intestinal enzymes work best in an alkaline environment pH 8.

12. **(D)** Different enzymes have different requirements. Gastric (stomach) enzymes work in the stomach where the pH is very acidic, a pH of 3. Intestinal enzymes work best in an alkaline environment, pH 8.

13. **(C)** Each villus contains capillaries that absorb amino acids, vitamins, and monosaccharides directly into the bloodstream and a lacteal, which absorbs fatty acids and glycerol into the lymphatic system.

14. **(D)** Gastrin, secretin, and CCK are hormones that regulate digestion. Gastrin is released by the stomach wall to stimulate the secretion of gastric juice. CCK, cholecystokinin, is released by the duodenum wall to stimulate the pancreas to release digestive enzymes and the gallbladder to release bile.

15. **(A)** Smooth muscles are also called involuntary muscles. They are under the control of the autonomic nervous system; we do not control these muscles consciously. Skeletal muscles are called striated or voluntary muscles. These are the muscles attached to bones that enable us to move. Skeletal muscles are under the control of the somatic system.

16. **(A)** Red blood cells carry hemoglobin and oxygen but very little carbon dioxide. Instead, most carbon dioxide is carried in the plasma as carbonic acid. Therefore, the more carbon dioxide there is in the blood, the more acidic the blood is. The other statements are correct.

17. **(B)** Leukocytes is the scientific name for white blood cells. Red blood cells are formed in the bone marrow. When they reach maturity and before they are released, the nucleus is removed so that there is more room for hemoglobin to carry more oxygen.

18. **(E)** The only artery carrying deoxygenated blood is the pulmonary artery, which carries blood to the lungs to drop off carbon dioxide and to pick up oxygen. All the other statements are correct.

19. **(B)** Erythrocytes are red blood cells. All other choices are involved with blood clotting.

20. **(A)** The vena cava carries blood from the body cells back to the heart into the right atrium. From there blood moves through the tricuspid valve to the right ventricle, where it is pumped to the lungs.

21. **(B)** Right atrium—right ventricle—pulmonary artery—lungs—pulmonary veins—left atrium—left ventricle

22. **(B)** Normal blood pressure for every adult is 120/80. The 120 is the systolic pressure and the 80 is the diastolic pressure. A person with blood pressure higher than that is unhealthy and suffering from hypertension. All the other statements are correct.

23. **(E)** Carbon dioxide is carried in the plasma. The bicarbonate ion is part of a buffering system that maintains the pH of the blood at 7.4. Leukocytes are white blood cells and fight disease. Erythrocytes are red blood cells and carry oxygen but very little carbon dioxide.

24. **(C)** Veins, not arteries, contain valves to help prevent the blood from flowing backward in the veins. All the other statements are correct.

25. **(C)** The islets of Langerhans in the pancreas release insulin, which lowers blood sugar levels.

26. **(D)** The anterior pituitary releases several hormones, including thyroid-stimulating hormone (TSH), human growth hormone (HGH), and ACTH, which stimulates the adrenal cortex to release corticosteroids

27. **(B)** The posterior pituitary produces oxytocin, which initiates labor, and ADH, antidiuretic hormone, which controls the reabsorption of water by the nephron of the kidney.

28. **(E)** The adrenal medulla releases adrenaline, the fight-or-flight hormone.

29. **(A)** The thyroid gland secretes thyroxin, which controls the rate of metabolism.

30. **(D)** Progesterone stimulates the thickening of the uterine lining that supports the development of an embryo. It is produced by the corpus luteum. (See the chapter, "Reproduction and Development.")

31. **(E)** HGH is human growth hormone. It is secreted by the anterior pituitary and stimulates growth.

32. **(A)** Glucagon is secreted by the islets of Langerhans of the pancreas. It causes the breakdown of glycogen and the increase in blood glucose levels.

33. **(C)** FSH is follicle-stimulating hormone, which is released by the anterior pituitary. In addition to stimulating the follicle in the ovaries in the female, it also stimulates the testes in the male.

34. **(B)** Insulin is produced in the islets of Langerhans of the pancreas and lowers blood glucose levels. It is lacking or insufficient in people with diabetes.

35. **(B)** Adrenaline (epinephrine) is the flight-or-fight hormone, not parathormone. Parathormone is secreted by the parathyroid and raises blood calcium levels. Calcium levels are necessary for normal blood clotting.

36. **(A)** Glucagon is produced by the islets of Langerhans in the pancreas, not the anterior pituitary. It raises blood glucose levels.

37. **(B)** Antidiuretic hormone (ADH) targets the collecting tubule in the nephron. It is produced in the hypothalamus and stored and released by the posterior pituitary. It increases the permeability of the collecting tubule to water so that more water can be reabsorbed back into the body and urine volume is reduced. When ADH levels are high, a more concentrated urine is excreted. When ADH production is decreased, more urine is excreted.

38. **(E)** Examples of tropic hormones are FSH, ACTH, and TSH.

39. **(A)** Filtration occurs by diffusion and is passive and nonselective. The filtrate contains everything small enough to diffuse out of the glomerulus and into Bowman's capsule, including glucose, salts, vitamins, wastes such as urea, and other small molecules.

40. **(B)** Urea is produced in the liver by the deamination of amino acids.

41. **(C)** Leukocytes are white bloods cells. They fight germs. The pH of the blood is consistently 7.4. Erythrocytes are red blood cells and carry oxygen. Type O blood is recessive, not dominant. The human circulatory system is a closed system.

42. **(E)** Antidiuretic hormone from the posterior pituitary targets the collecting tubule in the nephron. The thymus gland is important in the fetus for the proper development of the immune system. The lacteal is located inside the villi in the small intestine and absorbs fatty acids and glycerol. The glomerulus and Bowman's capsule are the site of filtration in the nephron and are not affected by ADH.

43. **(C)** The sympathetic nervous system is associated with fight-or-flight. The parasympathetic system is calming and relaxing, not excitatory.

44. **(B)** The parasympathetic system is generally calming, not excitatory. When you are in an excitatory mode, digestion ceases.

45. **(D)** Smooth muscle, like cardiac tissue, is involuntary muscle and under control of the autonomic nervous system.

Reproduction and Development

CHAPTER 16

- Introduction
- Asexual reproduction
- Sexual reproduction
- Menstrual cycle
- Spermatogenesis
- Oogenesis
- Embryonic development

Most animals show definite cycles of reproductive activity, often related to changing seasons. The periodic nature of this process allows animals to conserve resources and reproduce when environmental conditions favor the survival of offspring. Reproductive cycles are controlled by a combination of hormonal and environmental cues.

Animals may reproduce only asexually, only sexually, or alternate between the two. Asexual reproduction results in offspring that are genetically identical to the parents. This is an advantage when environmental conditions are stable.

Some eggs can develop by **parthenogenesis**, a process in which the egg develops without being fertilized and the adults that result are monoploid. This is characteristic of honeybees, where monoploid individuals are male drones and diploid individuals are female workers. Some sessile animals are **hermaphrodites** and can mate with any animal of their species. Both animals act as male and female, and both donate and receive sperm.

During sexual reproduction, a small flagellated monoploid sperm (n) fertilizes a larger, nonmotile monoploid egg (n) to form a diploid ($2n$) zygote. The zygote then undergoes **cleavage**, **gastrulation**, and **organogenesis**. Sexually reproducing fish and amphibians carry out external fertilization, where the female sheds thousands of eggs to be fertilized by sperm directly in the environment. The likelihood that sperm and egg will actually fuse is low; and the rate of predation of those that actually form a zygote is high. To compensate, millions of eggs and sperm are released at one time. Fish and amphibians reproduce this way. Birds, reptiles, and mammals carry out internal fertilization. Usually they produce fewer zygotes and provide more parental care.

Although there are many exceptions, Table 16.1 gives you the general idea.

TABLE 16.1

Types of Fertilization and Development

Animal	Fertilization	Development	Number of Eggs	Parental Care
Fish	External	External	Many	None
Amphibian	External	External	Many	None
Reptiles	Internal	External (inside the egg)	Few	Some
Birds	Internal	External (inside the egg)	Few	Much
Mammals	Internal	Internal	Few	Much

ASEXUAL REPRODUCTION

Asexual reproduction produces offspring genetically identical to the parent. It has several advantages over sexual reproduction.

1. It enables animals living in isolation to reproduce without a mate.
2. It creates numerous offspring quickly.
3. There is no expenditure of energy maintaining elaborate reproductive systems or hormonal cycles.
4. Because offspring are clones of the parent, asexual reproduction is advantageous when the environment is stable and favorable

TYPES OF ASEXUAL REPRODUCTION IN SAMPLE ORGANISMS

- **Fission** is the separation of an organism into two new cells. (Amoeba, bacteria)
- **Budding** involves the splitting off of new individuals from existing ones. (Hydra)
- **Fragmentation** and regeneration occur when a single parent breaks into parts that regenerate into new individuals. (Sponges, planaria, sea star)
- **Parthenogenesis** involves the development of an egg without fertilization. The resulting adult is monoploid or haploid. (Honeybees, some lizards)

SEXUAL REPRODUCTION

Sexual reproduction has one major advantage over asexual reproduction: variation. Each offspring is the product of both parents and may be better able to survive than either parent, especially in an environment that is changing.

The Human Male Reproductive System

Figure 16.1 illustrates the male reproductive system.

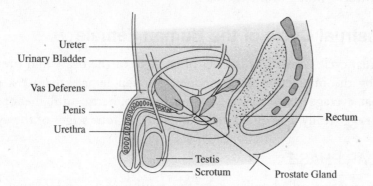

Figure 16.1

- **Testes** (testis, singular)—male gonads; the site of sperm formation in the testes
- **Vas deferens**—the duct that carries sperm during ejaculation from the epididymis to the penis
- **Prostate gland**—the large gland that secretes semen directly into the urethra
- **Scrotum**—the sac outside the abdominal cavity that holds the testes; the cooler temperature there enables sperm to survive
- **Urethra**—the tube that carries semen and urine

The Human Female Reproductive System

Figure 16.2 illustrates the female reproductive system.

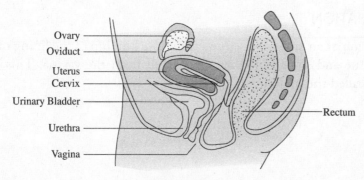

Figure 16.2

- **Ovary**—where meiosis occurs and where the secondary oocyte forms prior to birth
- **Oviduct or Fallopian tube**—where fertilization occurs; after ovulation, the egg moves through the oviduct to the uterus
- **Uterus**—where the blastula stage of the embryo will implant and develop during the nine-month gestation, should fertilization occur
- **Vagina**—the birth canal; during labor and delivery, the baby passes through the cervix and into the vagina
- **Cervix**—the mouth of the uterus
- **Endometrium**—lining of the uterus

The Menstrual Cycle of the Human Female

The menstrual cycle consists of a series of changes in the ovary and uterus that is controlled by the interaction of hormones. Human females release a gamete at intervals that average about every 28 days from puberty until menopause. The release of an egg (really a secondary oocyte) is one of four stages of the cycle.

FOLLICULAR PHASE

Several tiny cavities called follicles in the ovaries grow and secrete increasing amounts of estrogens in response to follicle-stimulating hormone (FSH) from the anterior pituitary.

OVULATION

The secondary oocyte ruptures out of the ovaries in response to a rapid increase in luteinizing hormone (LH) from the anterior pituitary. Ovulation occurs on or about the 14th day after menstruation.

LUTEAL PHASE

After ovulation, the corpus luteum (the cavity of the follicle left behind) forms and secretes estrogen and progesterone that thicken the endometrium (lining) of the uterus.

MENSTRUATION

If implantation of an embryo does not occur, the buildup of the lining of the uterus is shed. Tissue and some blood are discharged from the vagina. This bleeding is commonly called the period.

Hormonal Control of the Menstrual Cycle

The **hypothalamus** in the brain releases GnRH, which stimulates the anterior pituitary to release FSH and LH, which stimulate the ovary to release estrogen and progesterone. These two hormones prepare the uterus for implantation of an embryo.

Hypothalamus

Releases

↓

Gonadotropic-Releasing Hormone (GnRH)

↓

Stimulates

Anterior Pituitary

Releases

*Follicle-Stimulating Hormone
(FSH)*   *Luteinizing Hormone
(LH)*

Stimulates

↓

Ovary

Releases

Estrogen   *Progesterone*

Stimulates

↓   ↓

Thickening of the lining of the uterus

Spermatogenesis

Spermatogenesis, the process of sperm production, is a continuous process that begins at puberty and can continue into old age. It begins as luteinizing hormone (LH) induces the testes to produce testosterone. Together FSH and testosterone stimulate sperm production in the testes.

Each spermatogonium cell (2*n*) divides by *mitosis* to produce two primary spermatocytes (2*n*) which can each undergo *meiosis I* to produce two secondary spermatocytes (*n*). Each secondary spermatocyte then undergoes *meiosis II*, which yields four spermatids (*n*). These spermatids differentiate and move to the epididymis where they become motile. Note that each spermatogonium cell undergoes meiosis to produce four active, equal-sized sperm. Figure 16.3 shows the development of sperm by meiosis.

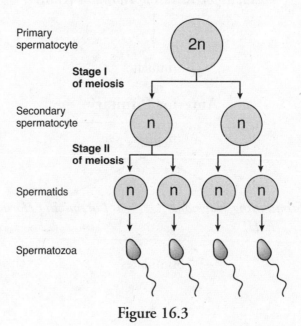

Figure 16.3

Oogenesis

Oogenesis, the production of ova, begins prior to birth. A female baby is born with all the primary oocytes she will ever have. Within the embryo, an oogonium cell (2*n*) undergoes *mitosis* to produce two primary oocytes (2*n*). These remain inactive within follicles in the ovaries until puberty, when they become reactivated by hormones. At that time, *meiosis I* occurs, producing **secondary oocytes** (*n*) that are released monthly at ovulation. *Meiosis II* does not occur until a sperm penetrates the secondary oocyte during fertilization. This could be 40 years after meiosis I.

During meiosis I and meiosis II the cytoplasm divides unequally. Almost all the cytoplasm remains in the egg, leaving two tiny polar bodies that have very little cytoplasm and which will disintegrate. Note that one primary oogonium cell produces only one active egg cell. Figure16.4 shows the development of an ovum by meiosis.

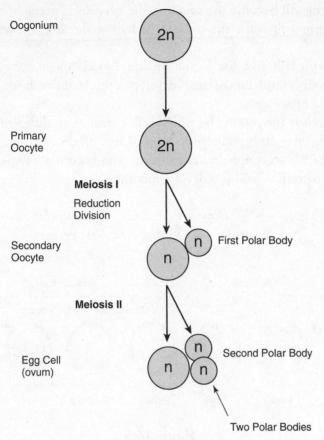

Figure 16.4

EMBRYONIC DEVELOPMENT

A small, flagellated, monoploid sperm (n) fertilizes a larger, nonmotile, monoploid egg (n) to form a diploid ($2n$) zygote. The zygote then undergoes cleavage, a succession of mitotic divisions that results in the formation of a hollow ball called a blastula.

Embryonic development consists of three stages: cleavage, gastrulation, and organogenesis.

Cleavage is the rapid mitotic cell division of the zygote that begins immediately after fertilization. The cells are dividing so quickly that individual cells have no time to grow in size. Embryologists arbitrarily consider the end of cleavage to be characterized by the production of a fluid-filled ball of cells called a blastula. The individual cells of the blastula are called blastomeres, and the fluid-filled center is a blastocoel.

Gastrulation is the continuation of the process that began during cleavage. It involves differentiation; the rearrangement of the blastula to produces a three-layered embryo called a gastrula. The gastrula consists of three differentiated layers called embryonic germ layers. These three germ layers are the **ectoderm**, **endoderm**, and **mesoderm**. They will develop into all the parts of the adult animal. See Figure 16.5.

- The **ectoderm** will become the skin and the nervous system.
- The **endoderm** will form the viscera, including the lungs, liver, and digestive organs.
- The **mesoderm** will give rise to the muscle, blood, and bones. Some primitive animals (sponges and cnidarians) develop a noncellular layer, the **mesoglea**, instead of the mesoderm.

Organogenesis is the process by which cells continue to differentiate, producing organs from the three embryonic germ layers. Once all the organ systems have been developed, the embryo simply increases in size and becomes a fetus.

Overall the pattern of embryo development is:

$$\text{Zygote} \rightarrow \textit{Cleavage} \rightarrow \text{Blastula} \rightarrow \textit{Gastrulation} \rightarrow$$
$$\text{Gastrula} \rightarrow \textit{Organogenesis} \rightarrow \text{Fetus}$$

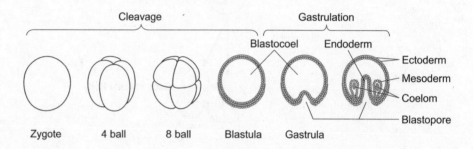

Figure 16.5

Extraembryonic Membranes of the Bird Embryo

You should know the structure and function of the four membranes that arise outside the bird embryo. They are known as the extraembryonic membranes. They consist of the chorion, yolk sac, amnion, and allantois.

1. **Chorion.** It lies under the shell and allows for diffusion of respiratory gases between the outside environment and the inside of the shell.

2. **Yolk sac.** It encloses the yolk, the food for the growing embryo.

3. **Amnion.** It encloses the embryo in protective amniotic fluid.

4. **Allantois.** It is analogous to the placenta in mammals. It is the conduit for respiratory gases to and from the embryo. It is also the place where the nitrogenous waste uric acid accumulates until the chick hatches.

Figure 16.6 illustrates the development of a zygote to a late embryo.

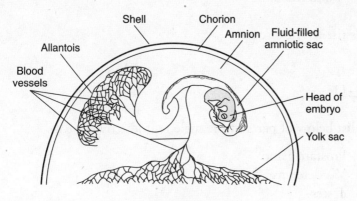

Figure 16.6

MULTIPLE-CHOICE QUESTIONS

1. Development of an embryo without fertilization is called

    (A) Budding
    (B) Parthenogenesis
    (C) Regeneration
    (D) Gastrulation
    (E) Organogenesis

2. All of the following are correct about asexual reproduction EXCEPT

    (A) it can create numerous offspring quickly
    (B) the offspring are genetically identical to the parent
    (C) it enables animals living in isolation to reproduce without a mate
    (D) there is no expenditure of energy maintaining reproductive systems
        or hormonal cycles
    (E) it is advantageous when the environment is changing and the
        organism is stressed

Questions 3–7

Refer to the list below of primary germ layers.

    (A) Ectoderm
    (B) Endoderm
    (C) Mesoderm

3. Gives rise to the lining of the digestive tract

4. Gives rise to the brain and eye

5. Gives rise to the blood

6. Gives rise to the bone

7. Gives rise to the viscera

Questions 8–11

Refer to the list below of parts of the male reproductive system.

   (A) Prostate gland
   (B) Vas deferens
   (C) Testes

8. Secretes semen directly into the urethra

9. Duct that carries sperm during ejaculation

10. Male gonads

11. Site of meiosis

Questions 12–15

Choose from the list below of parts of the female reproductive system.

   (A) Uterus
   (B) Cervix
   (C) Endometrium
   (D) Fallopian tube
   (E) Ovary

12. Where fertilization occurs

13. Where meiosis occurs

14. Lining of the uterus

15. Mouth of the uterus through which the baby passes

Questions 16–18

Refer to the terms below about the menstrual cycle.

   (A) Luteal Phase
   (B) Menstruation
   (C) Ovulation
   (D) Follicular phase

16. FSH stimulates the ovaries

17. Follicle secretes estrogen and progesterone

18. Secondary oocyte ruptures out of ovary

19. The end of cleavage is marked by the formation of

   (A) three embryonic layers
   (B) the archenteron
   (C) the blastula
   (D) zygote
   (E) the secondary oocyte

20. Gastrulation is marked by the formation of the

   (A) blastula
   (B) secondary oocyte
   (C) three embryonic layers
   (D) blastocoel
   (E) zygote

21. Which hormone is released by the anterior pituitary and directly causes ovulation?

   (A) Luteinizing hormone
   (B) Progesterone
   (C) Gonadotropic-releasing hormone (GnRH)
   (D) Follicle-stimulating hormone
   (E) Estrogen

22. Which hormone causes the lining of the uterus to thicken in preparation for implantation of the embryo?

   (A) Luteinizing hormone
   (B) Progesterone
   (C) Gonadotropic-releasing hormone (GnRH)
   (D) Follicle-stimulating hormone
   (E) Androgen

23. Which is the correct order of the stages of the menstrual cycle?

    (A) Follicular phase—ovulation—menstruation—luteal phase
    (B) Ovulation—luteal phase—menstruation—follicular phase
    (C) Follicular phase—ovulation—luteal phase—menstruation
    (D) Luteal phase—ovulation—menstruation—follicular phase
    (E) Luteal phase—follicular phase—ovulation—menstruation

Questions 24–25

Refer to the terms below that refer to the chick embryo.

    (A) Chorion
    (B) Yolk sac
    (C) Amnion
    (D) Allantois

24. Analogous to the placenta; where nitrogenous waste accumulates

25. Encloses the embryo in protective fluid

EXPLANATION OF ANSWERS

1. **(B)** Parthenogenesis is the development of an embryo without fertilization. Budding and regeneration are examples of asexual reproduction. Gastrulation is one stage in embryo development. Organogenesis is the formation of organs in an embryo.

2. **(E)** Offspring that result from asexual reproduction are identical to the parent. Therefore, asexual reproduction is advantageous when the environment is stable and unchanging.

3. **(B)** The endoderm gives rise to the viscera, the internal organs including the digestive tract.

4. **(A)** The ectoderm gives rise to the central nervous system (CNS) and the skin. The eye is considered part of the CNS.

5. **(C)** The mesoderm gives rise to the blood, bones, and muscle.

6. **(C)** The mesoderm gives rise to the blood, bones, and muscle.

7. **(B)** The endoderm gives rise to the viscera, the internal organs including the liver and kidneys.

8. **(A)** The prostate gland is a large gland that secretes semen directly into the urethra.

9. **(B)** The vas deferens is the duct that carries sperm from the testes to the penis.

10. **(C)** The testes are the sites of spermatogenesis.

11. **(C)** The site of sperm formation, meiosis, is the testes.

12. **(D)** The fallopian tube is also called the oviduct. It is the site of fertilization.

13. **(E)** Meiosis or oogenesis occurs in the ovaries of a fetus during the first trimester of embryonic development.

14. **(C)** The endometrium is the lining of the uterus. During the luteal phase, progesterone and estrogen are responsible for building up the endometrial lining.

15. **(B)** During labor, the cervix must open wide enough to allow the baby to pass through.

16. **(D)** FSH stands for follicle-stimulating hormone. It is released by the anterior pituitary and stimulates the maturation of several follicles in the ovaries.

17. **(A)** After ovulation, the corpus luteum (the cavity of the follicle left behind) forms and secretes estrogen and progesterone that thicken the endometrium (lining) of the uterus.

18. **(C)** The secondary oocyte ruptures out of the ovaries in response to a rapid increase in luteinizing hormone (LH) from the anterior pituitary. Ovulation occurs on or about the 14th day after menstruation.

19. **(C)** Cleavage is the rapid mitotic cell division of the zygote that begins immediately after fertilization. Embryologists arbitrarily consider the end of cleavage to be characterized by the production of a fluid-filled ball of cells called a blastula.

20. **(C)** Gastrulation involves the rearrangement of the blastula and produces a three-layered embryo called a gastrula. It consists of three differentiated layers called embryonic germ layers.

21. **(A)** Ovulation occurs in response to rapid increase in luteinizing hormone (LH) from the anterior pituitary. GnRH is released by the hypothalamus and stimulates the anterior pituitary to release FSH and LH. FSH stimulates follicles in the ovaries to mature a primary oocyte. Estrogen and progesterone are released by the corpus luteum and are responsible for thickening the uterine wall.

22. **(B)** Luteinizing hormone (LH) causes ovulation. GnRH is released by the hypothalamus and stimulates the anterior pituitary to release FSH and LH. FSH stimulates follicles in the ovaries to mature a primary oocyte. Androgens are male hormones like testosterone.

23. **(C)** During the follicular stage, follicles are stimulated. A secondary oocyte is released during ovulation. During the luteal phase, the corpus luteum begins to release progesterone and estrogen. If no implantation occurs in the lining of the uterus, the lining breaks down and is expelled from the body.

24. **(D)** In birds, the nitrogenous waste, uric acid, accumulates inside the shell in the allantois.

25. **(C)** The amnion encloses the amniotic fluid.

The Human Immune System

CHAPTER 17

- Introduction
- Nonspecific defense
- Specific defense
- Types of immunity—passive and active
- ABO blood type
- AIDS virus
- Allergies, antibiotics, and autoimmune disesease

We live in a sea of germs. They are in the air we breathe, the food we eat, and the water we drink and swim in. Diseases are spread in a number of different ways. Some are spread from person to person by direct physical contact or by sneezing or coughing. Some, like amoebic dysentery, are spread through contaminated water or food. Still others, like malaria or Lyme disease, are spread by infected animals.

The human body has evolved a complex system of defenses to fight **pathogens** (organisms that cause disease) and keep us healthy. Some of these defenses are like walls of a fortress. Others are like soldiers that engage in hand-to-hand combat once the invaders have crossed the moat and entered the fortress. There are also cells that act like warriors, shooting projectiles at the invaders who have breached the walls of the fortress.

One way in which the immune system defends the body is in a nonspecific way, by attacking anything foreign. The immune system can also identify certain invading cells and attack them specifically. Here are the basics of the immune system.

NONSPECIFIC DEFENSES

The nonspecific immune system consists of two lines of defense.

First Line of Defense

The first line of nonspecific defense is a barrier that helps prevent pathogens from entering the body. The body has several different types of barriers:

- Skin that blocks pathogens
- Mucous membranes that release mucus and trap microbes
- Cilia in the respiratory system that sweep out mucus with its trapped microbes
- Stomach acid that kills germs that enter through the nose and mouth

Second Line of Defense

Microbes that get into the body encounter the second line of nonspecific defense. It is meant to limit the spread of invaders in advance of specific immune responses. There are three types.

1. **Inflammatory response.** It is characterized by swelling, redness, soreness, and increased temperature in the area. The purpose of this process is to increase the blood supply to the area, thus increasing nutrients—including oxygen, and white blood cells to fight disease. The inflammatory response works in several ways.

    *   **Histamine** triggers vasodilation (enlargement of blood vessels), which increases blood supply to the area, bringing more phagocytes to gobble up germs. Histamine is also responsible for the symptoms of the common cold: sneezing, coughing, redness, and itching and runny nose and eyes—all an attempt to rid the body of invaders.
    *   Increased body temperature speeds up the immune system and makes it more difficult for microbes to function.

2. **Phagocytes.** These gobble up invading microbes. Macrophages ("giant eaters") are a type of white blood cell that extend **pseudopods** and engulf huge numbers of microbes over a long period of time.

3. **Interferons.** These chemicals are released by the immune system to block against viral infections.

SPECIFIC DEFENSES—THIRD LINE OF DEFENSE

The third line of defense is specific and consists of lymphocytes. There are two types of lymphocytes, B lymphocytes and T lymphocytes. Both originate in the bone marrow. Once mature, both cell types circulate in the blood, lymph, and lymphatic tissue: spleen, lymph nodes, tonsils, and adenoids. Both recognize different specific **antigens** (germs) (Strictly speaking, an antigen is anything that triggers an immune response).

1. **B lymphocytes.** These produce **antibodies** against a specific antigen in what is called a *humoral response.*

2. **T lymphocytes.** These fight pathogens by hand-to-hand combat in what is called a *cell-mediated response.*

Antibodies

Antibodies are part of the third line of defense—the specific immune response. Each antibody has the ability to bind to only one particular antigen. For example, antibodies against influenza bind to and neutralize only influenza virus; they have no effect on the polio virus. Antibodies neutralize antigens by binding to them and by forming an antigen-antibody complex that can then be gobbled up by a phagocyte.

Clonal Selection

Clonal selection is a fundamental mechanism in the development of immunity. Antigens that have entered the body bind to specific B or T lymphocytes. Once a lymphocyte has been selected, it becomes very metabolically active, proliferates (clones thousands of copies of itself), and differentiates into plasma cells and memory cells.

PLASMA CELLS

These fight antigens immediately in what is called the primary immune response. They do not live long.

MEMORY CELLS

These fight the same antigens that plasma cells do, but they remain circulating in the blood in small numbers for a lifetime. You have memory cells circulating in your blood that are specific for every viral infection you have ever been ill with and against every disease against which you have been vaccinated. You have memory cells specific for mumps, measles, rubella, polio, and so on. The capacity of the immune system to generate a secondary immune response is called immunological memory. The immunological memory is the mechanism that prevents you from getting any specific viral infection, such as chicken pox, more than once.

TYPES OF IMMUNITY

Passive immunity is temporary.

- Antibodies are borrowed and do not survive for long.
- Examples are maternal antibodies that pass through the placenta to the developing fetus or that pass through breast milk to the baby. The first milk that a newborn receives from mother is called colostrum and is 100 percent antibodies.

Active immunity is permanent.

- You make the antibodies yourself.
- An individual makes his or her own antibodies after being ill and recovering or after being given an immunization or vaccine. A vaccine contains either dead or live viruses or enough of the outercoat of a virus to stimulate a full immune response and to impart lifelong immunity.

ABO Blood Types

ABO antibodies circulate in the plasma of the blood and bind with ABO antigens in the event of an improper transfusion. Certain danger from a transfusion comes when *the recipient has antibodies to the donor's antigens*. However, before someone receives a transfusion of blood, samples of the recipient's and the donor's blood must be mixed in the lab to determine and ensure compatibility. This is called a cross-match.

> **STUDY TIP**
>
> Memorize which transfusions are acceptable, which are not, and why.

Blood type O is known as the universal donor because it has no blood cell antigens to be clumped by the recipient's blood. Blood type AB is known as the universal recipient because there are no antibodies to clump the donor's blood. Table 17.1 shows the antigens and antibodies present in each blood type.

TABLE 17.1

Blood Types

Blood Type	Antigens Present on the Surface of the RBCs	Antibodies Present Circulating in the Plasma
A	A	B (antibodies against B)
B	B	A (antibodies against A)
O	None	A and B (antibodies against A and B)
AB	A and B	No antibodies against A or B

AIDS Virus

AIDS stands for acquired immune deficiency syndrome. People with AIDS are highly susceptible to opportunistic diseases, infections, and cancers that take advantage of a collapsed immune system. The virus that causes AIDS, HIV (human immunodeficiency virus) mainly attacks helper T cells. HIV is a retrovirus. Once inside a cell, it reverse transcribes itself. That means that the viral RNA uses the enzyme reverse transcriptase to make DNA. This is the opposite of the typical DNA transcribing mRNA. The host cell then integrates this newly formed DNA into its own genome.

OTHER INFORMATION RELATED TO IMMUNITY

- **Allergies** are hypersensitive immune responses to certain substances called allergens. They involve the release of excessive amounts of histamine, an anti-inflammatory agent, which causes blood vessels to dilate. A normal allergic reaction involves redness, runny nose, and itchy eyes. Taking antihistamines can normally can counteract these symptoms. However, sometimes an acute allergic response can result in a life-threatening response called anaphylactic shock that can result in death within minutes.

- **Antibiotics** are medicines that kill bacteria or fungi. Although vaccines are given to prevent illness caused by viruses, antibiotics are administered after a person is sick. They cure the disease.

- **Vaccines prevent** viral infections. There is no treatment for viral infections, like there is for bacterial infections. New vaccines and treatments are always being developed. In fact, a new vaccine against a virus that causes cervical cancer has recently been developed. However, the best method for fighting any disease is to prevent it.

REMEMBER

Antibiotics cure.

REMEMBER

Vaccines prevent.

• **Autoimmune diseases** such as multiple sclerosis, lupus, arthritis, and juvenile diabetes are caused by a terrible mistake of the immune system. The system cannot properly distinguish between self and nonself. Instead, it perceives certain structures in the body as nonself and attacks them. In the case of multiple sclerosis, the immune system attacks the myelin sheath surrounding certain neurons.

MULTIPLE-CHOICE QUESTIONS

1. All of the following are part of the first line of defense of the immune system EXCEPT

   (A) leukocytes
   (B) skin
   (C) stomach acid
   (D) cilia
   (E) mucus

2. All of the following are true of the second line of defense EXCEPT

   (A) increased production of histamine
   (B) sneezing, redness, and itchy and runny nose and eyes
   (C) inflammatory response
   (D) stomach acid
   (E) phagocytes

3. All of the following are correct about the immune system EXCEPT

   (A) the first line of defense is nonspecific
   (B) the second line of defense is nonspecific
   (C) release of histamine is responsible for the inflammatory response
   (D) macrophages engulf huge numbers of microbes
   (E) T lymphocytes release antibodies

4. Which of the following is NOT part of the lymphatic system?

   (A) Tonsils
   (B) Spleen
   (C) Liver
   (D) Adenoids
   (E) Lymph nodes

5. Which of the following is true about histamine?

   (A) It is part of the body's first line of defense.
   (B) It kills germs by dissolving them.
   (C) It is a toxin released by microbes or germs.
   (D) It causes sneezing and a runny nose in an attempt to rid the body of germs.
   (E) It is part of the body's specific immune response.

Questions 6–10

Refer to terms below.

(A) T lymphocytes
(B) B lymphocytes
(C) Macrophages
(D) Antibodies
(E) Histamine

6. Produce antibodies

7. Fight pathogens by hand-to-hand combat

8. Neutralize specific antigens

9. Use pseudopods to engulf large numbers of germs

10. Attacks and kills infected body cells

11. Certain danger in a blood transfusion comes when the

(A) recipient has antigens to the donor blood
(B) recipient has antibodies to the donor blood
(C) donor has antibodies to the recipient's blood
(D) donor has antigens to the recipient's antigens
(E) donor's blood contains gamma globulin

12. A person with blood type A has

(A) A antibodies circulating in the plasma
(B) A antigens on the surface of the red blood cells
(C) B antibodies on the surface of the red blood cells
(D) A antigens circulating in the plasma
(E) O antibodies circulating in the plasma

13. A person with AB blood type has

(A) both A and B antigens on the surface of the red blood cells
(B) both A and B antigens circulating in the plasma
(C) O antigens circulating in the plasma
(D) no antigens on the surface of the red blood cells
(E) no antibodies on the surface of the red blood cells

14. Which is an example of passive immunity?

    (A) It is lifelong.
    (B) Babies who are nursing receive antibodies from their mothers.
    (C) You become resistant to a viral infection once you have recovered
        from it.
    (D) You become resistant to mumps after receiving the mumps vaccine.
    (E) People who have AIDS have antibodies against the virus but are still
        gravely ill.

15. Which of the following is CORRECT about the immune system?

    (A) Vaccines can cure certain common viral infections.
    (B) Allergies can be cured by antibiotics.
    (C) Multiple sclerosis is caused by an allergy.
    (D) Arthritis is an autoimmune disease.
    (E) Antibiotics can prevent many diseases.

16. All of the following are autoimmune diseases EXCEPT

    (A) arthritis
    (B) lupus
    (C) multiple sclerosis
    (D) AIDS
    (E) juvenile diabetes

17. Reverse transcriptase is an enzyme found in

    (A) macrophages
    (B) T lymphocytes
    (C) B lymphocytes
    (D) HIV virus
    (E) antibodies

18. Antibodies are

    (A) responsible for raising body temperature when you are ill
    (B) memory cells
    (C) plasma cells
    (D) specific
    (E) part of the first line of defense

19. Which is true about plasma cells?

    (A) When activated, they turn into memory cells.
    (B) They offer permanent resistance to disease.
    (C) They are produced from activated T cells; whereas memory cells
        are produced by activated B cells.
    (D) They fight antigens and die in battle.
    (E) They are found only in people who are infected with the
        AIDS virus.

20. Vasodilation

    (A) is triggered by histamine
    (B) means engulfing antigens
    (C) is caused by interferons
    (D) is part of an immunoglobin
    (E) is an important part of the specific immune response

EXPLANATION OF ANSWERS

1. **(A)** The first line of defense includes skin, stomach acid, mucous membranes that secrete mucus to trap germs, and cilia that sweep germs out of the body. Leukocytes are white blood cells, which are part of the second line of defense. All of the choices, however, are the immune system's nonspecific strategies.

2. **(D)** Stomach acid is a way of preventing germs from getting into the body. It is part of the first line of defense. The other choices are all part of the second line of defense. All choices are part of the immune system's nonspecific strategies.

3. **(E)** T lymphocytes are part of the cell-mediated response. It is the B lymphocytes that release antibodies as part of what is called the humoral response.

4. **(C)** The liver has many functions, including recycling of red blood cells, detoxification of poisons, and production of lipids, including cholesterol. However, it is not part of the lymphatic system. The lymphatic system consists of vessels and nodes and is a separate system in the body. White blood cells proliferate in lymph nodes, which is why your glands (lymph nodes) swell when you are fighting an infection.

5. **(D)** Histamine is part of the body's nonspecific and second line of defense. It is part of the inflammatory response, a means of increasing blood supply to increase the presence of nutrients and blood cells to help fight disease. It causes sneezing and coughing in an attempt to rid the body of invaders.

6. **(B)** B lymphocytes respond to infection by producing antibodies at an amazing rate.

7. **(A)** T lymphocytes kill our own body cells that have been infected with viruses, or other pathogens by attacking them directly.

8. **(D)** Antibodies neutralize antigens.

9. **(C)** Macrophages ("giant eaters") extend pseudopods and engulf huge numbers of microbes nonspecifically over a long period of time. They are an important part of the second line of defense.

10. **(A)** T lymphocytes kill our own body cells that have been infected with viruses, or other pathogens by attacking them directly.

11. **(B)** Certain danger in a blood transfusion comes when the recipient has antibodies to the donor's blood. Transfusion reactions can be mild or severe and fatal. Type O blood is the universal donor (in theory) because it has no antigens on the surface of its red blood cells. Likewise, type AB blood is the universal recipient (in theory) because it has no antibodies circulating in its plasma. However, doing a blood transfusion is not theoretical, and a cross-match must be done first to determine the compatibility of the blood.

12. **(B)** ABO antigens are found on the surface of red blood cells. A and B antibodies circulate in the plasma. There are no O antigens or antibodies. O is like zero; it means none.

13. **(A)** A person with AB blood has both A and B antigens on the surface of the red blood cells. This is an example of codominance, where both genes and traits are expressed. ABO antigens are found on the surface of red blood cells. A and B antibodies circulate in the plasma. There are no O antigens or antibodies. O is like zero; it means none.

14. **(B)** Passive immunity is borrowed immunity. Someone else makes the antibodies. It is temporary. The first milk that a newborn receives from the mother is called colostrum and is 100 percent antibodies. Choice (A) is a characteristic of active immunity. Choices (C) and (D) are examples of active immunity. About choice (E): people who have AIDS do produce antibodies against the HIV virus, and that is an example of active immunity. Unfortunately, the antibodies are ineffective against the disease.

15. **(D)** Multiple sclerosis, lupus, juvenile diabetes, and arthritis are all autoimmune diseases. They are caused by an error in the immune system. The immune system attacks and kills the body's cells. In the case of multiple sclerosis, the immune system attacks and kills neurons in the central nervous system. Vaccines can prevent viral infections, such as colds, flu, and polio. Antibiotics do not prevent disease; they can cure bacterial infections such as strep throat.

16. **(D)** AIDS is caused by the HIV virus. All the rest are autoimmune diseases. Autoimmune diseases are caused by an error in the immune system. The immune system attacks and kills the body's cells. In the case of multiple sclerosis, the immune system attacks and kills neurons in the brain.

17. **(D)** The virus that causes AIDS, HIV (human immunodeficiency virus), mainly attacks helper T cells. HIV is a retrovirus. Once inside a cell, it reverse transcribes itself. That means that the viral RNA uses reverse transcriptase to transcribe in reverse and make DNA. The host cell then integrates this newly formed DNA into its own genome. It remains as a provirus in the host nucleus, directing the production of new viruses.

18. **(D)** Antibodies are specific. For every antigen, there exists a specific antibody. They are part of the third line of defense—the specific immune response.

19. **(D)** Plasma cells fight antigens immediately in what is called the primary immune response. They do not live long because they die in battle. Memory cells survive a lifetime and offer permanent resistance to disease. Both B cells and T cells proliferate, when triggered, into plasma and memory cells. Statement (E) is a false statement.

20. **(A)** Histamine triggers vasodilation (enlargement of blood vessels), which increases blood supply to the area, bringing more phagocytes to gobble up germs. It is part of the second line of defense and is nonspecific. Interferons are chemicals released by the immune system that block against viral infections. Immunoglobins are antibodies.

Animal Behavior

• Introduction	• Learning
• Fixed action pattern	• Social behavior

An organism's behavior is important for its survival and for the successful production of offspring. The study of behavior and its relationship to its evolutionary origins is called ethology. Foremost in the field of ethology are three scientists who shared the Nobel Prize in 1973: Karl von Frisch, Konrad Lorenz, and Niko Tinbergen. Von Frisch is known for his extensive studies of honeybee communication and his famous description of the bee waggle dance. Niko Tinbergen is known for his elucidation of the **fixed action pattern**, and Konrad Lorenz is famous for his work with **imprinting**. Here are some basic concepts in the field of animal behavior.

FIXED ACTION PATTERN

A fixed action pattern (FAP) is an innate, *highly stereotypical behavior* that, once begun, is continued to completion no matter how useless or silly looking. FAPs are initiated by external stimuli called **sign stimuli**. When these stimuli are exchanged between members of the same species, they are known as **releasers**. An example of a FAP studied by Tinbergen involves the stickleback fish, which attacks other males that invade its territory. The releaser for the attack is the red belly of the intruder. The stickleback will not attack an invading male stickleback lacking a red underbelly, but it will readily attack a nonfishlike wooden model as long as a splash of red is visible.

LEARNING

Learning is a sophisticated process in which the responses of the organism are modified as a result of experience. The capacity to learn can be tied to length of life and complexity of the brain. If the animal has a very short life span, like a fruit fly, it has no time to learn, even if it has the ability. It must therefore rely on fixed action patterns. In contrast, if the animal lives a long time and has a complex brain, then a large part of its behavior is dependent on prior experience and learning.

Habituation

Habituation is one of the simplest forms of learning in which an animal comes to ignore a persistent stimulus so it can go about its business. If you tap the dish containing a hydra, it will quickly shrink and become immobile. If you keep tapping, after a while the hydra will begin to ignore the tapping, elongate, and continue moving about. It has become habituated to the stimulus.

Associative Learning

Associative learning is one type of learning in which one stimulus becomes linked to another through experience. Examples of associative learning are **classical conditioning** and **operant conditioning**.

CLASSICAL CONDITIONING

Classical conditioning, a type of associative learning, is widely accepted because of the ingenious work of Ivan Pavlov in the 1920s. Normally, dogs salivate when exposed to food. Pavlov trained dogs to associate the sound of a bell with food. The result of this conditioning was that dogs would salivate upon merely hearing the sound of the bell even though no food was present.

OPERANT CONDITIONING

Operant conditioning, also called trial and error learning, is another type of associative learning. An animal learns to associate one of its own behaviors with a reward or punishment and then repeats or avoids that behavior. The best-known studies involving operant conditioning were done by B. F. Skinner in the 1930s. In one study, a rat was placed into a cage containing a lever that released a pellet of food. At first, the rat would depress the lever only by accident and would receive food as a reward. The rat soon learned to associate the lever with the food and would depress the lever at will. Similarly, an animal can learn to carry out a behavior to avoid punishment. Such systems of rewards and punishment are the basis of most animal training.

Imprinting

Imprinting is learning that occurs during a sensitive or critical period in the early life of an individual and is irreversible for the length of that period. When you see ducklings following closely behind their mother, you are seeing the result of successful imprinting. Mother-offspring bonding in animals that depend on parental care is critical to the safety and development of the offspring. If the pair does not bond, the parent will not care for the offspring and the offspring will die. At the end of the juvenile period, when the offspring can survive without the parent, the response disappears.

Classic imprinting experiments were carried out by Konrad Lorenz with geese. Geese hatchlings will follow the first thing they see that moves. Although the object is usually the mother goose, it can be a box tied to a string or in the case of the classic experiment, it was Konrad Lorenz himself. Lorenz was the first thing the hatchlings saw and they became imprinted on the scientist. Wherever he went, they followed.

SOCIAL BEHAVIOR

Social behavior is any kind of interaction between two or more animals, usually of the same species. It is a relatively new field of study, only developed in the 1960s. Types of social behaviors are cooperation, agonistic, dominance hierarchies, territoriality, and altruism.

Cooperation

Cooperation enables the individuals to carry out a behavior, such as hunting, which they can do as a group more successfully than they can do separately. Lions or wild dogs will hunt in a pack, enabling them to bring down an animal larger than an individual could ever bring down alone.

Agonistic Behavior

Agonistic behavior is aggressive behavior. It involves a variety of threats or actual combat to settle disputes between individuals. These disputes are commonly over access to food, mates, or shelter. It involves both real aggressive behavior as well as ritualistic or symbolic behavior. One combatant does not have to kill the other. The use of symbolic behavior often prevents serious harm. A dog shows aggression by baring its teeth and erecting its ears and hair. It stands upright to appear taller and looks directly at its opponent. If the aggressor succeeds in scaring the opponent, the loser engages in submissive behavior that says, "You win, I give up." Examples of submissive behaviors are looking down or away from the winner. Submissive dogs or wolves put their tail between their legs and run off. Once two individuals have settled a dispute and established their relationship by agonistic behavior, future encounters between them usually do not involve combat or posturing.

Dominance Hierarchies

Dominance hierarchies are pecking order behaviors that dictate the social position of an animal in a culture. This is commonly seen in hens where the alpha animal (top-ranked) controls the behaviors of all the others. The next in line, the beta animal, controls all others except the alpha animal. Each animal threatens all animals beneath it in the pecking order. The top-ranked animal is assured of first choice of any resource, including food after a kill, the best territory, or the most-fit mate.

Territoriality

A **territory** is an area an organism defends and from which other members of the community are excluded. Territories are established and defended by agonistic behaviors. They are used for capturing food, mating, and rearing young. The size of the territory varies with its function and the amount of resources available.

Altruism

Altruism is a behavior that reduces an individual's reproductive fitness (the animal may die) while increasing the fitness of the group or family. When a worker honeybee stings an intruder in defense of the hive, the worker bee usually dies. However, it increases the fitness of the queen bee that lays all the eggs. How can altruism evolve if the altruistic individual dies? The answer is called kin selection. When an individual sacrifices itself for the family, it is sacrificing itself for relatives (the kin), which share similar genes. The kin are selected as the recipients of the altruistic behavior. They are saved and can pass on their genes. Altruism evolved because it increases the number of copies of a gene common to a related group.

MULTIPLE-CHOICE QUESTIONS

Questions 1–7

Choose from the list of scientists below.

    (A) Niko Tinbergen
    (B) Karl von Frisch
    (C) B. F. Skinner
    (D) Konrad Lorenz
    (E) Ivan Pavlov

1. Described the waggle dance in honeybees

2. Imprinting

3. Trained dogs to salivate at the sound of a bell

4. Classical conditioning

5. Taught rats in cages to depress a lever to release food

6. Baby geese followed him everywhere

7. Explained fixed action pattern

Questions 8–12

Choose from the terms below.

    (A) Fixed action pattern
    (B) Habituation
    (C) Classical conditioning
    (D) Imprinting
    (E) Operant conditioning

8. Innate, highly stereotypical behavior that must continue until it is completed

9. Trial and error learning

10. Sequence of behaviors that is unchangeable and carried to completion once initiated

11. Initially, the amoeba moved away from the strong light; but after a while, it resumed its normal movement pattern

12. This is the way dogs are trained

13. A sophisticated process in which the responses of the organism are modified as a result of experience is called

    (A) fixed action pattern
    (B) habituation
    (C) imprinting
    (D) classical conditioning
    (E) learning

14. This behavior reduces an individual's reproductive fitness while increasing the fitness of the family.

    (A) Altruism
    (B) Agonistic behavior
    (C) Territoriality
    (D) Cooperation
    (E) Imprinting

15. You want to train your puppy to wait at the curb until you tell him to cross the road. Your friend advises you to give your dog a treat every time he does as you ask. Your friend is advising that you train the dog using

    (A) operant conditioning
    (B) classical conditioning
    (C) imprinting
    (D) fixed action pattern
    (E) habituation

16. _____ is learning that occurs during a sensitive or critical period in early life and is irreversible for the length of the period.

    (A) Habituation
    (B) Operant conditioning
    (C) Trial and error learning
    (D) Imprinting
    (E) Classical conditioning

17. "Mary had a little lamb; its fleece was white as snow. And everywhere that Mary went, the lamb was sure to go." The behavior of the lamb is best described as

    (A) habituation
    (B) imprinting
    (C) operant conditioning
    (D) classical conditioning
    (E) fixed action pattern

18. Fixed action patterns are initiated by external stimuli called

    (A) fixed action pattern
    (B) sign stimuli
    (C) agonistic behavior
    (D) dominance hierarchies
    (E) sensitive periods

19. Animals that help other animals are expected to be

    (A) stronger than other animals
    (B) related to the animals they help
    (C) male
    (D) female
    (E) disabled in some way

20. An animal that sacrifices itself for its relatives is exhibiting

    (A) operant conditioning
    (B) kin selection
    (C) classical conditioning
    (D) imprinting
    (E) habituation

EXPLANATION OF ANSWERS

1. **(B)** Karl von Frisch is known for his extensive studies of honeybee communication and his famous description of the waggle dance in bees.

2. **(D)** Konrad Lorenz carried out imprinting experiments. When you see ducklings following closely behind their mother, you are seeing the result of successful imprinting.

3. **(E)** In the 1920s, Ivan Pavlov conditioned dogs to salivate at the sound of a bell. This is known as classical conditioning.

4. **(E)** An example of classical conditioning is the training of dogs to salivate at the sound of a bell.

5. **(C)** B. F. Skinner is the scientist whose name is linked with operant conditioning.

6. **(D)** A goose hatchling follows the first thing it sees moving after it hatches. In the famous case, goose hatchlings saw and imprinted on Lorenz, not the mother goose.

7. **(A)** Tinbergen worked with stickleback fish, demonstrating highly stereotypical behavior known as fixed action pattern.

8. **(A)** This is the definition of fixed action pattern

9. **(E)** Learning by trial and error is also called operant conditioning. B. F. Skinner carried out the famous studies in operant conditioning.

10. **(A)** Fixed action pattern is highly stereotypical.

11. **(B)** Habituation is one of the simplest forms of learning. An animal comes to ignore a persistent stimulus so it can go about its business.

12. **(E)** A system of rewards and punishment, as described in operant conditioning, is the way most animals are trained.

13. **(E)** The capacity to learn is tied to the length of an animal's life and the complexity of the brain. If the animal has a short life span, like an insect, it has no time to learn and must rely on fixed action patterns.

14. **(A)** Altruism enhances the fitness of the family group although it may mean sacrificing the individual.

15. **(A)** Operant conditioning is trial and error conditioning. This is how we train animals.

16. **(D)** Konrad Lorenz made this theory of imprinting and himself famous.

17. **(B)** This describes the geese that followed Konrad Lorenz around everywhere. He was the first moving object they saw after hatching, so they imprinted on him.

18. **(B)** Sign stimuli are external stimuli that trigger fixed action patterns. Those exchanged between members of the same species are known as releasers.

19. **(B)** Social behavior is any kind of interaction among two or more animals, usually of the same species. Altruism is a behavior that will help the other members of the family group, who have many genes in common.

20. **(B)** Altruism is a behavior that ultimately increases the success of the related group. In kin selection, an individual sacrifices itself for the family, a group that shares its genes. Perhaps the individual that sacrificed itself will not get to pass on its genes, but its relatives will.

Ecology

CHAPTER 19

- Vocabulary
- Properties of populations
- Population growth
- Community structure and population interactions
- The food chain
- Ecological succession
- Biomes
- Chemical cycles—water, carbon, and nitrogen
- Humans and the biosphere

Ecology is the study of the interactions of organisms with their physical environment and with each other. Here is some basic vocabulary for the topic.

1. A **population** is a group of individuals of one species living in one area who can interbreed and interact with each other.
2. A **community** consists of all the organisms living in one area.
3. An **ecosystem** includes all the organisms in a given area as well as the abiotic (nonliving) factors with which they interact.
4. **Abiotic factors** are nonliving and include temperature, water, sunlight, wind, rocks, and soil.
5. **Biotic factors** include all the organisms with which an organism might react, such as birds, insects, predators, prey, and parasites.
6. The **biosphere** is the global ecosystem.

PROPERTIES OF POPULATIONS

Populations are defined by their size, density, and dispersion.

Size

Size is the total number of individuals in a population. Four variables limit the size of a population: the number of births, the number of deaths, immigration, and emigration.

Density

Density is the number of individuals per unit area or volume. It is often very difficult, if not impossible, to count the number of organisms inhabiting a certain area. For example, imagine trying to count the number of ants in 1 acre of land. Instead, scientists use sampling techniques to estimate the number of organisms living in one area. One sampling technique commonly used to estimate the size of a population is called mark and recapture. In this technique, organisms are captured, tagged, and then released. Some time later, the same process is repeated and a special mathematical formula is used to determine the density of the population.

Dispersion

Dispersion is the pattern of spacing of individuals within the area the population inhabits. The most common pattern of dispersion is clumped. Fish travel this way in schools because there is safety in numbers. Some populations are spread in a uniform pattern. For example, certain plants may secrete toxins that keep away other plants that would compete for limited resources. Random spacing occurs in the absence of any special attractions or repulsions. Trees can be spaced randomly in a forest. Figure 19.1 shows different dispersion patterns.

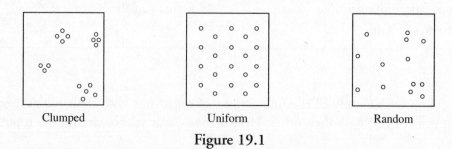

Clumped   Uniform   Random

Figure 19.1

POPULATION GROWTH

Every population has a characteristic **biotic potential**, *the maximum rate at which a population could increase under ideal conditions.* Different populations have different biotic potentials that are influenced by several factors. These factors include:

- Age at which reproduction begins
- Life span during which the organisms are capable of reproducing
- Number of reproductive periods in the lifetime
- Number of offspring the organism is capable of reproducing

Regardless of whether a population has a large or small biotic potential, certain characteristics about growth are common to all organisms.

Figure 19.2 shows a graph of classic population growth. After an initial period of slow growth when the organism becomes accustomed to the new environment, the population explodes and grows exponentially. The population grows until it reaches the maximum that the environment can support (**carrying capacity**). After some undetermined amount of time, the population crashes. Many factors can cause a population to crash: predation, parasitism, severe competition, an end to resources, and/or too much waste that poisons the environment.

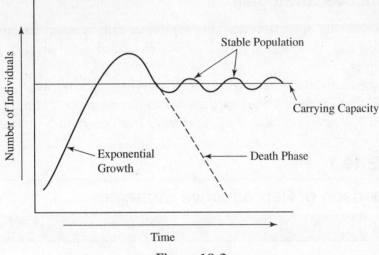

Figure 19.2

Exponential Growth

The simplest model for population growth is one with unrestrained or exponential growth—a population with no predation, parasitism, or competition, no immigration or emigration, and in an environment with unlimited resources. This is characteristic of a population that has been recently introduced into an area, such as a sample of bacteria newly inoculated onto a petri dish. Although exponential growth is usually short-lived in nature, the human population has been in the exponential growth phase for over 300 years.

Carrying Capacity

Ultimately, there is a limit to the number of individuals that can occupy one area at a particular time. That limit is called the carrying capacity (K). Each particular environment has its own carrying capacity around which the population size oscillates.

In addition, the carrying capacity can change as the environmental conditions change. This is illustrated in Figure 19.3.

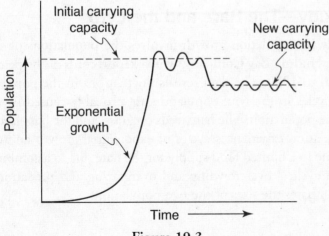

Figure 19.3

Reproductive Strategies

Some organisms are opportunistic. They reproduce rapidly when the environment is uncrowded and resources are vast. They are referred to as r-strategists. Insects are good examples.

Other organisms, the K-strategists (K for carrying capacity), tend to maximize population size near the carrying capacity for an environment. Mammals are good examples of K-strategists. Table 19.1 compares r-strategists and K-strategists.

TABLE 19.1

Comparison of Reproductive Strategies	
r-strategists	**K-strategists**
Many, small young	Few, large young
Little or no parenting	Intensive parenting
Rapid maturation	Slow maturation
Reproduce once	Reproduce many times
Example: insects	Example: mammals

Limiting Factors

Limiting factors are those factors that limit population growth. They are divided into two categories: density-dependent and density-independent factors.

- Density-dependent factors are those factors that increase directly as the population density increases. They include competition for food, buildup of wastes, predation, and disease.
- Density-independent factors are those factors whose occurrence is unrelated to the population density. They include earthquakes, storms, and naturally occurring fires and floods.

A Case Study—The Hare and the Lynx

A perfect study in population growth involves the populations of snowshoe hare and lynx at the Hudson Bay Company, which kept records of the pelts sold by trappers from 1850 to 1930. The data reveals fluctuations in the populations of both animals. The cycles in the lynx population are probably caused by cyclic fluctuations in the hare population. The hare feeds on grass, and the lynx feeds on the hare. The hare population experiences cycles of exponential growth and crashes. These are probably due to a limited food supply for the hare due to a combination of malnutrition from cyclical overcrowding and overgrazing and predation by the lynx. Figure 19.4 compares the sizes of the two populations.

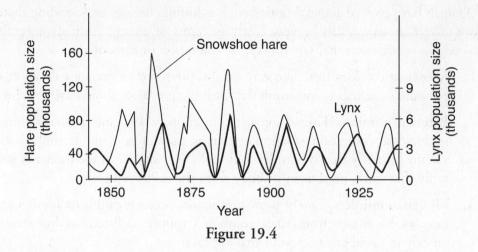

Figure 19.4

COMMUNITY STRUCTURE AND POPULATION INTERACTIONS

Communities are made of populations that interact with the environment and with each other. These interactions are very complex but can be divided into five categories: competition, predation, **parasitism**, **mutualism**, and **commensalism**.

Competition

The Russian scientist, G. F. Gause, developed the competitive exclusion principle after studying the effects of competition within a species in a laboratory setting. He worked with two very similar species, *Paramecium caudatum* and *Paramecium aurelia*. When he cultured them separately, each population grew rapidly and then leveled off at the carrying capacity. When he put the two cultures together, *P. aurelia* had the advantage and drove the other species to extinction. His principle states that *two species cannot coexist in a community if they share a niche, that is, if they use the same resources.*

In nature, there are two possible outcomes besides extinction if two species inhabit the same area and occupy the same niche. One of the species will become extinct as happened in Gause's experiment, or one will evolve through natural selection to exploit different resources. This second process is called resource partitioning. Another possibility is what occurred in the Galapagos Islands, where finches evolved different beak sizes and were able to eat different kinds of seeds and avoid competition. This divergence in adaptation is called character displacement.

Predation

Predation can refer to one animal eating another animal, or it can also refer to animals eating plants. For their protection, animals and plants have evolved defenses against predation.

Plants have evolved spines and thorns and chemical poisons such as strychnine, mescaline, morphine, and nicotine to fend off attack by animals.

Animals have evolved active defenses, such as hiding, fleeing, or defending themselves. These, however, can be very costly in terms of energy. Animals have also evolved passive defenses that rely on cryptic coloration or camouflage.

1. **Aposematic coloration.** The very bright, often red or orange coloration of poisonous animals is a warning that possible predators should avoid them.

2. **Batesian mimicry.** This is copycat coloration, where one harmless animal mimics the coloration of another that is poisonous. One example is the viceroy butterfly that is harmless but looks very similar to the monarch butterfly that stores poisons in its body from the milkweed plant.

3. **Müllerian mimicry.** Two or more poisonous species resemble each other and gain an advantage from their combined numbers. Predators learn more quickly to avoid any prey with that appearance.

Mutualism

Mutualism is a symbiotic relationship where both organisms benefit (+/+). An example is the bacteria that live in the human intestine and produce vitamins for the person.

Commensalism

Commensalism is a symbiotic relationship in which one organism benefits and one is neither helped nor harmed by the other organism (+/o). Barnacles that attach themselves to the underside of a whale benefit by gaining access to a variety of food sources as the whale swims into different areas. The whale is unaware of the barnacle.

Parasitism

Parasitism is a symbiotic relationship (+/−) where one organism, the parasite, benefits while the host is harmed. A tapeworm in the human intestine is an example.

THE FOOD CHAIN

The **food chain** is the pathway along which food is transferred from one trophic or feeding level to another. Energy, in the form of food, moves from the producers to the herbivores to the carnivores. Only about 10 percent of the energy stored in any trophic level is converted to organic matter at the next trophic level. This means that if you begin with 1,000 g of plant matter, the food chain can support 100 g of herbivores (primary consumer), 10 g of secondary consumer (carnivore), and only 1 g of tertiary consumer (carnivore). As a result of the loss of energy from one trophic level to the next, food chains never have more than four or five trophic levels. A good model to demonstrate the interaction of the organisms in the food chain and the loss of energy is the food pyramid shown in Figure 19.5.

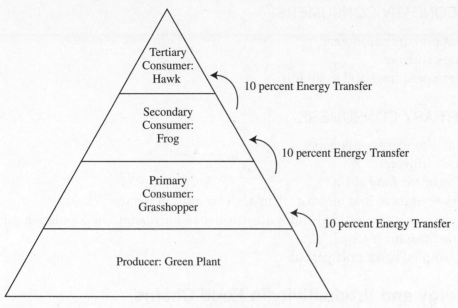

Figure 19.5

Food chains are not isolated; they are interwoven with other food chains into a food web. An animal can occupy one trophic level in one food chain and a different trophic level in another food chain. Humans, for example, can be primary consumers when eating vegetables but are tertiary consumers when eating a steak. Here are two examples of food chains, one terrestrial and one marine, each with four trophic levels.

Producers → Primary Consumers → Secondary Consumers → Tertiary Consumers

Terrestrial Food Chain

Green Plant → Grasshopper → Frog → Hawk

Marine Food Chain

Phytoplankton → Zooplankton → Small Fish → Shark

> **STUDY TIP**
>
> Diatoms and phytoplankton are producers.

PRODUCERS

Producers convert light energy to chemical bond energy and have the greatest biomass of any trophic level. They include green plants, **diatoms**, and **phytoplankton**. Diatoms are photosynthetic protists that drift in the oceans. Phytoplankton are algae and photosynthetic bacteria that drift passively in aquatic environments. Diatoms and phytoplankton are the basis for most marine and freshwater aquatic ecosystems.

PRIMARY CONSUMERS

- Eat producers
- Are herbivores
- Examples: grasshoppers and zooplankton (microscopic arthropods)

SECONDARY CONSUMERS

- Eat primary consumers
- Are carnivores
- Examples: frogs and small fish

TERTIARY CONSUMERS

- Eat secondary consumers
- Are carnivores
- Top of the food chain
- Fewer of these (less biomass) than any organism in the food chain
- Least stable trophic level and most sensitive to fluctuations in populations of the other trophic levels
- Examples: hawk or larger fish

Energy and Productivity in Food Chains

When ecologists study energy transformations within ecosystems, they discuss productivity. **Gross primary productivity** is the amount of energy converted to chemical energy by photosynthesis per unit time in an ecosystem. **Net primary productivity** is the gross primary productivity minus the energy used by the primary producers for respiration.

Biological Magnification

Organisms at higher trophic levels have greater concentration of accumulated toxins stored in their bodies than those at lower trophic levels. This is called **biological magnification**. The bald eagle almost became extinct because Americans sprayed heavily with the pesticide DDT in the 1950s. DDT entered the food chain and accumulated in the bald eagle, at the top of the food chain. Because DDT interferes with the deposition of calcium in eggshells, the thin-shelled eggs were broken easily and few eaglets hatched. DDT is now outlawed, and the bald eagle was saved from extinction.

Decomposers

Decomposers—bacteria and fungi—are usually not depicted in any diagram of a food chain. However, without decomposers to recycle nutrients back to the soil to nourish plants, there would be no food chain and no life.

ECOLOGICAL SUCCESSION

Most communities are not stable. They are dynamic, always changing. The size of a population increases and decreases around the carrying capacity. Migration of a new species into a habitat can alter the entire food chain. Major disturbances, both natural and human-made, like volcanic eruptions, strip mining, clear cutting a forest, and forest fires, can suddenly and drastically destroy a community or an entire ecosystem. What follows this destruction is the process of sequential rebuilding of the ecosystem called ecological succession.

If the rebuilding begins in a lifeless area where even soil has been removed, the process is called primary ecological succession. *The essential and dominant characteristic of primary ecological succession is soil building.* After an ecosystem is destroyed, the first organisms to inhabit a barren area are pioneer organisms like lichens (a symbiont consisting of algae and fungi) and mosses, which are introduced into the area as spores by the wind. Soil develops gradually as rocks weather and organic matter accumulates from the decomposed remains of the pioneer organisms. Once soil is present, pioneer organisms are overrun by other, larger organisms: grasses, bushes, and then trees. The final stable community that remains is called the climax community. It remains until the ecosystem is once again destroyed by a blowout, a disaster that destroys the ecosystem.

One example of primary succession that was studied in detail is at the southern edge of Lake Michigan. As the lakeshore gradually receded northward after the last ice age (10,000 years ago), it left a series of new beaches and sand dunes exposed. Now, someone who begins at the water's edge and walks away from the water for several miles will pass through a series of communities that were formed in the last 10,000 years. These communities represent the various stages, beginning with bare, sandy beach and ending with a climax community of old well-established forests. In some cases, the climax community is a beech-sugar maple forest. In other areas, the forest is a mix of hickory and oak.

The process known as secondary succession occurs when an existing community has been cleared by some disturbance that leaves the soil intact. This can be seen in the 1988 fires in Yellowstone National Park that destroyed all the old growth, that had been dominated by lodge pole pine, but left the soil intact. Within one year, the burned areas in Yellowstone were covered with new vegetation.

BIOMES

Biomes are very large regions of Earth whose distribution depends on the amount of rainfall and the temperature in an area. Each biome is characterized by different vegetation and animal life. There are many biomes, including freshwater, marine, and terrestrial biomes. In the northern hemisphere, from the equator to the most northerly climes, there is a trend in terrestrial biomes: from tropical rain forest, to desert, to grasslands, to temperate deciduous forest, to taiga, and finally to tundra in the north. Changes in altitude produce effects similar to changes in latitudes. On the slopes of the Appalachian Mountains in the east and the Rockies and coastal ranges in the west, there is a similar trend in biomes. As elevation increases and temperatures and humidity decrease, one passes through temperate deciduous forest to taiga to tundra. Here is an overview of the major biomes of the world.

Marine

- The largest biome, covering three-fourths of Earth's surface
- The most stable biome, with temperatures that vary little because water has a high heat capacity and there is such an enormous volume of water
- Provides most of Earth's food and oxygen
- Subdivided into different regions classified by amount of sunlight they receive, distance from shore and water depth, and whether open water or ocean bottom

STUDY TIP

The marine biome is the largest and most stable on Earth.

Tropical Rain Forest

- This biome is found near the equator with abundant rainfall, stable temperatures, and high humidity.
- Although these forests cover only 4 percent of Earth's land surface, they account for more than 20 percent of Earth's net carbon fixation (food production).
- It has the most plant species diversity of any biome on Earth. It may have as many as 50 times the number of species of trees as does a temperate forest.
- Dominant trees are very tall with interlacing tops that form a dense canopy, keeping the floor of the forest dimly lit even at midday. The canopy also prevents rain from falling directly onto the forest floor, but leaves drip rain constantly.
- Many trees are covered with epiphytes, photosynthetic plants that grow on other trees rather than supporting themselves. They are not parasites but may kill the trees inadvertently by blocking the light.
- This biome has the most animal species diversity of any biome and includes birds, reptiles, mammals, and amphibians.

Desert

- It receives less than 10 inches of rainfall per year; not even grasses can survive.
- A desert experiences the most extreme temperature fluctuations of any biome. Daytime *surface* temperatures can be as high as 70°C. With no moderating influence of vegetation, heat is lost rapidly at night. Shortly after sundown temperatures drop drastically.
- Characteristic plants are the drought-resistant cacti with shallow roots to capture as much rain as possible during hard and short rains that are characteristic of the desert.
- Other plants include sagebrush, creosote bush, and mesquite.
- There are many small annual plants that are stimulated to grow only after a hard rain. They germinate, send up shoots and flowers, and die all within a few weeks.
- Most animals are active at night or during a brief early-morning period or late afternoon, when the heat is not so intense. During the day, animals remain cool by burrowing underground or hiding in the shade.
- Cacti can expand to hold extra water and have modified leaves called spines, which protect against animals attacking a cactus for its water.
- As an example of how severe conditions in a desert can be, in the Sahara desert, there are regions of hundreds of miles across that are completely barren of any vegetation.
- Characteristic animals include rodents, kangaroo rats, snakes, lizards, arachnids, insects, and a few birds.

Temperate Grasslands

- Cover huge areas in both the temperate and tropical regions of the world
- Characterized by low total annual rainfall or uneven seasonal occurrence of rainfall, making conditions inhospitable for forests
- Principal grazing mammals include bison and pronghorn antelope in the United States, and wildebeest and gazelle in Africa
- Burrowing mammals, such as prairie dogs and other rodents, are common

Temperate Deciduous Forest or Boreal Forest

- Found in the northeast of North America, south of the taiga, and characterized by trees that drop their leaves in winter
- Includes many more plant species than does the taiga
- Shows *vertical stratification* of plants and animals—some species live on the ground, some in the low branches, and some in the treetops
- Rich soil due to decomposition of leaf litter
- Principal mammals include squirrels, deer, foxes, and bears that are dormant or hibernate through the cold winter

Conifer Forest–Taiga

- Found in northern Canada and much of the world's northern regions
- Dominated by conifer (evergreens) forests, like spruce and fir
- Landscape is dotted with lakes, ponds, and bogs
- Has very cold winters
- Is the largest terrestrial biome
- Characterized by heavy snowfall; trees are shaped with branches directed downward to prevent heavy accumulations of snow from breaking their branches
- Principal large mammals include moose, black bear lynx, elk, wolverines, martens, and porcupines
- Flying insects and birds are prevalent in summer
- Has greater variety in animal species than does the tundra

> **STUDY TIP**
>
> The taiga—the conifer forest—is the largest terrestrial biome.

Tundra

- Located in the far northern parts of North America, Europe, and Asia
- Called the permafrost, permanently frozen subsoil found in the farthest point north, including Alaska
- Commonly referred to as the frozen desert because it gets very little rainfall, which cannot penetrate the frozen ground
- Has the appearance of gently rolling plains with many lakes, ponds, and bogs in depressions
- Insects, particularly flies, are abundant
- Vast numbers of birds nest in the tundra in the summer to eat the insects and migrate south in the winter
- Principal mammals include reindeer, caribou, Arctic wolves, Arctic foxes, Arctic hares, lemmings, and polar bears
- Though the number of individual organisms in the tundra is high, the number of species is small

CHEMICAL CYCLES

Although Earth receives a constant supply of energy from the sun, chemicals must be recycled. You must know several chemical cycles: the carbon, nitrogen, and water cycles.

The Water Cycle

Water evaporates from the Earth, forms clouds, and rains over the oceans and land. Some rain percolates through the soil and makes its way back to the seas. Some evaporates directly from the land, but most evaporates from plants by transpiration.

The Carbon Cycle

The basis of this is the reciprocal processes of photosynthesis and respiration.

- Cell respiration by animals and bacterial decomposers adds carbon dioxide to the air and removes oxygen.
- The burning of fossil fuels adds carbon dioxide to the air.
- Photosynthesis removes carbon dioxide from the air and adds oxygen.

The Nitrogen Cycle

Very little nitrogen enters ecosystems directly from the air. Most of it enters ecosystems by way of bacterial processes.

- Nitrogen-fixing bacteria live in the nodules in the roots of legumes and convert free nitrogen (N_2) into the ammonium ion (NH_4^+).
- Nitrifying bacteria convert the ammonium ion (NH_4^+) into nitrites (NO_2^-) and then into nitrates (NO_3^-).
- Denitrifying bacteria convert nitrates (NO_3^-) into free atmospheric nitrogen (N_2).

HUMANS AND THE BIOSPHERE

As the human population has grown in size, we have intruded upon and altered or destroyed many ecosystems. We are responsible for the deforestation of millions of acres of land and the destruction of vast wetlands. We have caused groundwater contamination and depletion, the elimination of habitats, and the loss of biodiversity. Humans threaten to make the Earth uninhabitable as our population increases exponentially and as we waste natural resources and pollute the air and water. Here are some specific examples of how humans have altered Earth's ecosystem.

Eutrophication of the Lakes

We have disrupted freshwater ecosystems, causing a process called eutrophication. Runoff from sewage and manure from pastures increase nutrients in lakes and cause excessive growth of algae and other plants. Shallow areas become choked with weeds, and swimming and boating become impossible. As large populations of photosynthetic organisms die, two things happen. First, organic material accumulates on the lake bottom, and reduces the depth of the lake, second, detritivores use up oxygen as they decompose the dead organic matter. Lower oxygen levels make it impossible for some fish to live. As more fish die, decomposers expand their activity and oxygen levels continue to decrease. The process continues, more organisms die, the oxygen levels decrease, more decomposing matter accumulates on the lake bottom, and ultimately, the lake disappears. Figure 19.6 shows the eutrophication of a lake over ten years.

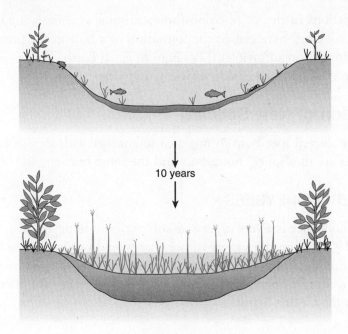

10 years

Figure 19.6

Acid Rain

Acid rain is caused by pollutants in the air from the combustion of fossil fuels. Nitrogen and sulfur pollutants in the air turn into nitric, nitrous, sulfurous, and sulfuric acids, which cause the pH of the rain to be less than 5.6. This causes the death of the organisms in lakes and damages ancient stone architecture.

Toxins

Toxins from industry have gotten into the food chain. Most cattle and chicken feeds contain antibiotics and hormones to accelerate animal growth but may have serious ill effects on humans who eat the chicken and beef. Any carcinogens or teratogens (causing birth defects) that get into the food chain accumulate and remain in our body's fatty tissues.

Global Warming

Excessive burning of fossil fuels has caused the concentrations of carbon dioxide in the air to increase to such high levels that it causes the **greenhouse effect**. This means that carbon dioxide and water vapor in the air absorb much of the infrared radiation reflecting off of Earth, causing the average temperature on Earth to rise. This increase in temperature is called **global warming** and could have disastrous effects for the world's population. An increase of 1.0°C on average temperature worldwide would cause the *polar ice caps to melt* raising the level of the seas. Eventually, major coastal cities in the United States, including New York, Los Angeles, and Miami would be under water.

Depleting the Ozone Layer

The accumulations in the air of **chlorofluorocarbons**, chemicals used for refrigerants and aerosol cans, have caused the formation of a hole in the protective **ozone layer**. This allows more ultraviolet (UV) light to reach Earth. This is responsible for an increase in the incidence of **skin cancer** worldwide.

Introduction of New Species

Humans have moved species from one area to another with serious consequences. Two examples are the "killer" honeybees and the zebra mussel.

THE "KILLER" HONEYBEES

The African honeybee is a very aggressive subspecies of honeybee that was brought to Brazil in 1956 to breed a variety of bee that would produce more honey in the tropics than the Italian honeybee. The African honeybees escaped by accident and have been spreading throughout the Americas. By the year 2000, these bees killed ten people in the United States.

THE ZEBRA MUSSEL

In 1988, the zebra mussel, a fingernail-sized mollusk, native to Asia, was discovered in a lake near Detroit. No one knows how the mussel got transplanted there. However, scientists assume that it was accidentally carried by a ship from a freshwater port in Europe to the Great Lakes. Without any local natural predator to limit its population, the mussel population exploded. They were first discovered when they were found to have clogged the water intake pipes of those cities whose water is supplied by Lake Erie. To date, the zebra mussel has caused millions of dollars of damage. In addition, the influx of the zebra mussel threatens several indigenous species with extinction by out-competing them.

Pesticides vs. Biological Control

Scientists have developed a variety of pesticides, chemicals that kill organisms that we consider to be undesirable. These include insecticides, herbicides, fungicides, and mice and rat killers. On the one hand, these pesticides save lives by increasing food production and by killing animals that carry and cause diseases like bubonic plague (diseased rats) and malaria (anopheles mosquitoes). On the other hand, exposure to pesticides can cause cancer in humans. Moreover, spraying with pesticides ensures the development of resistant strains of pests through natural selection. The pests come back stronger than before. This problem requires that we spray more and more, which means more people will be exposed to these toxic chemicals.

An alternative to widescale spraying with pesticides is called biological control. The following are some biological solutions to get rid of pests without using dangerous chemicals.

1. Use crop rotation—change the crop planted in a field.
2. Introduce natural enemies of the pests—you must be careful, however, that you do not disrupt a delicate ecological balance by introducing an invasive species.
3. Use natural plant toxins instead of synthetic ones.
4. Use insect birth control—male insect pests can be sterilized by exposing them to radiation and then releasing them into the environment to mate unsuccessfully with females.

MULTIPLE-CHOICE QUESTIONS

1. All the organisms of one species living in one area are known as a(n)

   (A) community
   (B) population
   (C) species
   (D) ecotype
   (E) detritivore

2. All the organisms living in one area are known as a(n)

   (A) community
   (B) population
   (C) species
   (D) ecotype
   (E) detrivore

Questions 3–8

Choose from the terms below.

   (A) Herbivore
   (B) Producer
   (C) Primary consumer
   (D) Secondary consumer
   (E) Decomposer

3. Converts solar energy to chemical bond energy

4. Has the greatest mass in any food chain

5. Fungi are an example of this

6. Contains the most pesticides of any organism on the list

7. A primary consumer is also this

8. Has the least biomass in a food chain

9. All of the following are correct about the food chain EXCEPT

   (A) bacteria and fungi act as decomposers that recycle nutrients
   (B) producers are always at the bottom of any food chain
   (C) energy is lost at each trophic level
   (D) food chains never have more than 4 or 5 trophic levels
   (E) pesticides tend to be concentrated at the producer level because producers have the largest mass in a food chain

Questions 10–18

Choose from the names of the biomes below.

   (A) Tropical rain forest
   (B) Taiga
   (C) Temperate grasslands
   (D) Marine
   (E) Tundra

10. Called the permafrost

11. Characteristic organisms: large mammals including black bear and elk, and conifer forests

12. Smallest number of species

13. Covers less than 4 percent of Earth's land surface but produces 20 percent of Earth's food

14. Largest biome

15. Most diversity of species

16. Rapidly decreasing due to human interaction

17. Most stable biome

18. Provides most of Earth's oxygen

19. Epiphytes are

   (A) a climax community of plants in the desert
   (B) photosynthetic plants that grow on trees rather than supporting themselves
   (C) vegetation found in grasslands
   (D) decomposers in the taiga
   (E) animals found in the tundra

20. Eutrophication refers to

    (A) the process that causes the depletion of the ozone layer
    (B) global warming
    (C) the process that happens to a lake that absorbs too many nutrients
    (D) the invasion of new species that causes damage to an ecosystem
    (E) the process whereby one species outcompetes another species

Questions 21–24

Choose from the terms below.

    (A) Mutualism
    (B) Parasitism
    (C) Commensalism
    (D) None of the above
    (E) A, B, and C

21. Both organisms benefit

22. One organism benefits; the other organism is not affected by the first organism

23. Example of symbiosis

24. Example of endosymbiosis

25. The result of the action of detritivores is

    (A) formation of ammonia
    (B) increase in transpiration
    (C) nitrogen fixation
    (D) decrease in oxygen
    (E) increase in oxygen

Questions 26–29

Choose from the terms below about the nitrogen cycle.

    (A) Nitrogen-fixing bacteria
    (B) Denitrifying bacteria
    (C) Nitrifying bacteria

26. These live in the roots of legumes

27. Convert the ammonium ion into nitrates

28. Convert nitrates into free nitrogen in the atmosphere

29. Convert free nitrogen from the atmosphere into the ammonium ion

30. Which of the following shows the correct order of the biomes as you move from the equator to the North Pole in the western hemisphere?

   (A) Tropical rain forest—desert—temperate rain forest—taiga—tundra
   (B) Tundra—taiga—temperate deciduous forest—desert—tropical rain forest
   (C) Desert—taiga—tundra—temperate deciduous forest—tropical rain forest
   (D) Tundra—temperate deciduous forest—taiga—desert—tropical rain forest
   (E) Desert—tropical rain forest—temperate deciduous forest—taiga—tundra

EXPLANATION OF ANSWERS

1. **(B)** A population is a group of individuals of one species living in one area who can interbreed and interact with each other. A community consists of all the organisms living in one area. A species is defined in terms of reproductive isolation. In order for organisms to belong to the same species, they must be a naturally occurring, interbreeding group that produces fertile offspring. Ecotypes refer to differences within a species determined by geography. The jackrabbit in the south has dark fur to blend in with the colors in the woods and long ears to radiate off excess heat in a warm climate. Jackrabbits in the cold north have white fur to blend in with snow and short ears close to the head to minimize heat loss.

2. **(A)** A community consists of all the organisms living in one area. A population is a group of individuals of one species living in one area who can interbreed and interact with each other. A species is defined in terms of reproductive isolation. In order for organisms to belong to the same species, they must be a naturally occurring, interbreeding group that produces fertile offspring. Ecotypes refer to differences within a species determined by geography. The jackrabbit in the south has dark fur to blend in with the colors in the woods and long ears to radiate off excess heat in a warm climate. Jackrabbits in the cold north have white fur to blend in with snow and short ears close to the head to minimize heat loss.

3. **(B)** Producers are all photosynthetic. They use energy from the sun to make glucose.

4. **(B)** The producer level in any food chain has more biomass than any other level. The higher up the food chain, the less the biomass.

5. **(E)** Fungi and bacteria decompose dead organic matter and recycle it in an ecosystem.

6. **(D)** Organisms at higher trophic levels have the greatest concentrations of accumulated toxins stored in their bodies.

7. **(A)** Herbivores are animals that only eat plants. Because of this, they are primary consumers.

8. **(D)** The higher up the food chain, the less the biomass.

9. **(E)** Pesticides become concentrated at the top of the food chain. Organisms at higher trophic levels absorb all the pesticides from all the organisms they eat in their lifetimes. The bald eagle was almost pushed to extinction after Americans sprayed heavily with the pesticide DDT in the 1950s.

10. **(E)** The tundra is found in the far northern parts of North America, Europe, and Asia. It is commonly referred to as the frozen desert because it gets very little rainfall that can penetrate the frozen ground.

11. **(B)** The taiga is located in northern Canada and much of the world's northern regions. It is dominated by conifers (evergreen) forests, like spruce and fir.

12. **(E)** Although the number of individual organisms is high, the number of species in the tundra is small.

13. **(A)** Although this forest covers only 4 percent of Earth's land surface, it contains the most biological diversity. It may have as many as 50 times the number of species of trees as a temperate forest.

14. **(D)** The marine biome is the largest biome, covering three-fourths of Earth's surface.

15. **(A)** The tropical rain forest has the most diverse species of animals on Earth. It may have 50 times the number of species of the temperate forest.

16. **(A)** Humans are clear-cutting vast expanses of tropical rain forest and destroying much of the habitat for hundreds of species of animals.

17. **(D)** The marine biome is the most stable biome (the least variation in temperature) and largest producer of oxygen on Earth.

18. **(D)** The marine biome is the most stable biome (the least variation in temperature) and largest producer of oxygen on Earth.

19. **(B)** Many trees are covered with epiphytes that are photosynthetic but that end up killing the plants they grow on because the epiphytes deprive them of light.

20. **(C)** Runoff from sewage and manure from pastures increase nutrients in lakes and cause excessive growth of algae and other plants. Shallow areas become choked with weeds, and swimming and boating become impossible. As these large populations of photosynthetic organisms die, two things happen. First, organic material accumulates on the lake bottom and reduces the depth of the lake. Second, detritivores use up oxygen as they decompose the dead organic matter.

21. **(A)** Mutualism is the type of symbiotic relationship where both organisms benefit. An example is the bacteria that live in our intestines and produce vitamins. Another is nitrogen fixing bacteria that live in nodules of the roots of legumes and convert free nitrogen into ammonium that the plant can use.

22. **(C)** Commensalism is the type of symbiosis where one organism benefits and one is not affected by the other organism. Barnacles that attach themselves to the underside of a whale benefit by gaining access to a variety of food sources as the whale swims into different areas. The whale is unaffected by the barnacles.

23. **(E)** Symbiosis is the close association and interdependence of two organisms. If one organism evolves, the other must evolve in order to survive.

24. **(D)** Endosymbiosis is the phenomenon that explains how mitochondria and chloroplasts evolved. They were once tiny, free-living protista that took up residence inside larger prokaryotes. The relationship was so beneficial to each one that the relationship became permanent.

25. **(D)** Detritivores use up oxygen as they decompose dead organic matter.

26. **(A)** Nitrogen-fixing bacteria live in the nodules on the roots of legumes like peanuts and convert free nitrogen from the air into ammonium, a nitrogen compound the plant can use.

27. **(C)** Nitrifying bacteria convert the ammonium ion into nitrites and then into nitrates.

28. **(B)** Denitrifying bacteria convert nitrates into free atmospheric nitrogen.

29. **(A)** Nitrogen-fixing bacteria live in the nodules in the roots of legumes and convert free nitrogen into the ammonium ion.

30. **(A)** The key to this answer is that the biome located across the globe at the equator is the tropical rain forest. In addition, tundra is found at the most northern reaches of the hemisphere across northern Alaska.

SAMPLE TESTS

Answer Sheet 1
SAMPLE TEST 1

1 Ⓐ Ⓑ Ⓒ Ⓓ Ⓔ 16 Ⓐ Ⓑ Ⓒ Ⓓ Ⓔ 31 Ⓐ Ⓑ Ⓒ Ⓓ Ⓔ 46 Ⓐ Ⓑ Ⓒ Ⓓ Ⓔ
2 Ⓐ Ⓑ Ⓒ Ⓓ Ⓔ 17 Ⓐ Ⓑ Ⓒ Ⓓ Ⓔ 32 Ⓐ Ⓑ Ⓒ Ⓓ Ⓔ 47 Ⓐ Ⓑ Ⓒ Ⓓ Ⓔ
3 Ⓐ Ⓑ Ⓒ Ⓓ Ⓔ 18 Ⓐ Ⓑ Ⓒ Ⓓ Ⓔ 33 Ⓐ Ⓑ Ⓒ Ⓓ Ⓔ 48 Ⓐ Ⓑ Ⓒ Ⓓ Ⓔ
4 Ⓐ Ⓑ Ⓒ Ⓓ Ⓔ 19 Ⓐ Ⓑ Ⓒ Ⓓ Ⓔ 34 Ⓐ Ⓑ Ⓒ Ⓓ Ⓔ 49 Ⓐ Ⓑ Ⓒ Ⓓ Ⓔ
5 Ⓐ Ⓑ Ⓒ Ⓓ Ⓔ 20 Ⓐ Ⓑ Ⓒ Ⓓ Ⓔ 35 Ⓐ Ⓑ Ⓒ Ⓓ Ⓔ 50 Ⓐ Ⓑ Ⓒ Ⓓ Ⓔ
6 Ⓐ Ⓑ Ⓒ Ⓓ Ⓔ 21 Ⓐ Ⓑ Ⓒ Ⓓ Ⓔ 36 Ⓐ Ⓑ Ⓒ Ⓓ Ⓔ 51 Ⓐ Ⓑ Ⓒ Ⓓ Ⓔ
7 Ⓐ Ⓑ Ⓒ Ⓓ Ⓔ 22 Ⓐ Ⓑ Ⓒ Ⓓ Ⓔ 37 Ⓐ Ⓑ Ⓒ Ⓓ Ⓔ 52 Ⓐ Ⓑ Ⓒ Ⓓ Ⓔ
8 Ⓐ Ⓑ Ⓒ Ⓓ Ⓔ 23 Ⓐ Ⓑ Ⓒ Ⓓ Ⓔ 38 Ⓐ Ⓑ Ⓒ Ⓓ Ⓔ 53 Ⓐ Ⓑ Ⓒ Ⓓ Ⓔ
9 Ⓐ Ⓑ Ⓒ Ⓓ Ⓔ 24 Ⓐ Ⓑ Ⓒ Ⓓ Ⓔ 39 Ⓐ Ⓑ Ⓒ Ⓓ Ⓔ 54 Ⓐ Ⓑ Ⓒ Ⓓ Ⓔ
10 Ⓐ Ⓑ Ⓒ Ⓓ Ⓔ 25 Ⓐ Ⓑ Ⓒ Ⓓ Ⓔ 40 Ⓐ Ⓑ Ⓒ Ⓓ Ⓔ 55 Ⓐ Ⓑ Ⓒ Ⓓ Ⓔ
11 Ⓐ Ⓑ Ⓒ Ⓓ Ⓔ 26 Ⓐ Ⓑ Ⓒ Ⓓ Ⓔ 41 Ⓐ Ⓑ Ⓒ Ⓓ Ⓔ 56 Ⓐ Ⓑ Ⓒ Ⓓ Ⓔ
12 Ⓐ Ⓑ Ⓒ Ⓓ Ⓔ 27 Ⓐ Ⓑ Ⓒ Ⓓ Ⓔ 42 Ⓐ Ⓑ Ⓒ Ⓓ Ⓔ 57 Ⓐ Ⓑ Ⓒ Ⓓ Ⓔ
13 Ⓐ Ⓑ Ⓒ Ⓓ Ⓔ 28 Ⓐ Ⓑ Ⓒ Ⓓ Ⓔ 43 Ⓐ Ⓑ Ⓒ Ⓓ Ⓔ 58 Ⓐ Ⓑ Ⓒ Ⓓ Ⓔ
14 Ⓐ Ⓑ Ⓒ Ⓓ Ⓔ 29 Ⓐ Ⓑ Ⓒ Ⓓ Ⓔ 44 Ⓐ Ⓑ Ⓒ Ⓓ Ⓔ 59 Ⓐ Ⓑ Ⓒ Ⓓ Ⓔ
15 Ⓐ Ⓑ Ⓒ Ⓓ Ⓔ 30 Ⓐ Ⓑ Ⓒ Ⓓ Ⓔ 45 Ⓐ Ⓑ Ⓒ Ⓓ Ⓔ 60 Ⓐ Ⓑ Ⓒ Ⓓ Ⓔ

E Section

61 Ⓐ Ⓑ Ⓒ Ⓓ Ⓔ 66 Ⓐ Ⓑ Ⓒ Ⓓ Ⓔ 71 Ⓐ Ⓑ Ⓒ Ⓓ Ⓔ 76 Ⓐ Ⓑ Ⓒ Ⓓ Ⓔ
62 Ⓐ Ⓑ Ⓒ Ⓓ Ⓔ 67 Ⓐ Ⓑ Ⓒ Ⓓ Ⓔ 72 Ⓐ Ⓑ Ⓒ Ⓓ Ⓔ 77 Ⓐ Ⓑ Ⓒ Ⓓ Ⓔ
63 Ⓐ Ⓑ Ⓒ Ⓓ Ⓔ 68 Ⓐ Ⓑ Ⓒ Ⓓ Ⓔ 73 Ⓐ Ⓑ Ⓒ Ⓓ Ⓔ 78 Ⓐ Ⓑ Ⓒ Ⓓ Ⓔ
64 Ⓐ Ⓑ Ⓒ Ⓓ Ⓔ 69 Ⓐ Ⓑ Ⓒ Ⓓ Ⓔ 74 Ⓐ Ⓑ Ⓒ Ⓓ Ⓔ 79 Ⓐ Ⓑ Ⓒ Ⓓ Ⓔ
65 Ⓐ Ⓑ Ⓒ Ⓓ Ⓔ 70 Ⓐ Ⓑ Ⓒ Ⓓ Ⓔ 75 Ⓐ Ⓑ Ⓒ Ⓓ Ⓔ 80 Ⓐ Ⓑ Ⓒ Ⓓ Ⓔ

M Section

81 Ⓐ Ⓑ Ⓒ Ⓓ Ⓔ 86 Ⓐ Ⓑ Ⓒ Ⓓ Ⓔ 91 Ⓐ Ⓑ Ⓒ Ⓓ Ⓔ 96 Ⓐ Ⓑ Ⓒ Ⓓ Ⓔ
82 Ⓐ Ⓑ Ⓒ Ⓓ Ⓔ 87 Ⓐ Ⓑ Ⓒ Ⓓ Ⓔ 92 Ⓐ Ⓑ Ⓒ Ⓓ Ⓔ 97 Ⓐ Ⓑ Ⓒ Ⓓ Ⓔ
83 Ⓐ Ⓑ Ⓒ Ⓓ Ⓔ 88 Ⓐ Ⓑ Ⓒ Ⓓ Ⓔ 93 Ⓐ Ⓑ Ⓒ Ⓓ Ⓔ 98 Ⓐ Ⓑ Ⓒ Ⓓ Ⓔ
84 Ⓐ Ⓑ Ⓒ Ⓓ Ⓔ 89 Ⓐ Ⓑ Ⓒ Ⓓ Ⓔ 94 Ⓐ Ⓑ Ⓒ Ⓓ Ⓔ 99 Ⓐ Ⓑ Ⓒ Ⓓ Ⓔ
85 Ⓐ Ⓑ Ⓒ Ⓓ Ⓔ 90 Ⓐ Ⓑ Ⓒ Ⓓ Ⓔ 95 Ⓐ Ⓑ Ⓒ Ⓓ Ⓔ 100 Ⓐ Ⓑ Ⓒ Ⓓ Ⓔ

Biology E/M Sample Test 1

With Answers and Analysis

CHAPTER **20**

Directions: Each question or incomplete statement is followed by five possible answers or completions. For both Biology-E and Biology-M, select the one choice that is the best answer and fill in the corresponding space on the answer sheet.

1. Sexually reproducing organisms show greater variation than asexually reproducing ones because

   (A) they exhibit fewer mutations
   (B) they exhibit a greater mutation rate
   (C) asexually reproducing organisms do not have internal membranes
   (D) of recombination of alleles
   (E) they are larger

2. Several species of rhododendron are growing in the same area. All of the plants are capable of hybridization, but none ever do because some of the plants produce pollen in early June while others produce pollen in late June. This best describes an evolutionary process known as

   (A) survival of the fittest
   (B) overpopulation
   (C) reproductive isolation
   (D) artificial selection
   (E) stabilizing selection

3. All of the following contribute to variation in a population EXCEPT

   (A) mutation
   (B) isolation
   (C) sexual reproduction
   (D) conjugation
   (E) genetic drift

4. Oxygen released by plants comes from

(A) air
(B) carbon dioxide
(C) glucose
(D) chlorophyll
(E) water

5. All of the following are mammals EXCEPT

(A) tiger
(B) ape
(C) kangaroo
(D) blue jay
(E) duck-billed platypus

6. Which of the following is a density-independent factor?

(A) Disease
(B) Famine
(C) Floods
(D) Predation
(E) Increase in toxins in the environment

7. A gene pool in a population of jackrabbits in a field remained constant for many generations. The most probable reason for this stable gene pool is that

(A) no migration occurred in a large population with random mating and no mutations
(B) no migration occurred in a small population with random mating and no mutations
(C) no migration occurred in a large population with nonrandom mating and no mutations
(D) there was much migration into and out of the large population, but mating was random and there were few mutations
(E) the population was small with no mutations, no migrations, and nonrandom mating

8. All of the following about plasma membrane structure and function are correct EXCEPT

(A) all plasma membranes have the identical composition and structure
(B) diffusion of gases across a membrane require that the membrane be moist
(C) facilitated diffusion is an example of passive transport
(D) proteins serve as membrane channels
(E) plasma membranes contain receptors that are specific for the molecules they uptake

9. Which of the following exhibits internal fertilization, external development of the embryo, few eggs, and much parenting?

   (A) Mammals
   (B) Amphibians
   (C) Reptiles
   (D) Birds
   (E) Fish

10. A solution with a pH of 5 is _____ times more acidic than a solution with a pH of 7.

   (A) 1/10
   (B) 1/100
   (C) 10
   (D) 100
   (E) 1,000

11. Vitamins are essential for normal cell function. They are important because they

   (A) function as an energy source
   (B) are hormones
   (C) directly assist in the normal conduction of impulses
   (D) resist pH changes
   (E) enable enzymes to function normally

12. Tendons connect _____ to _____; ligaments connect _____ to _____.

   (A) bone to bone; bone to muscle
   (B) bone to muscle; bone to bone
   (C) bone to bone; muscle to muscle
   (D) muscle to muscle; bone to bone
   (E) ligaments to bone; tendons to bones

13. Food chains never consist of more than 4 or 5 trophic levels. The reason for this is

   (A) energy is lost along the food chain
   (B) there are fewer primary consumers in the world than secondary consumers
   (C) producers are always shown on the bottom of any food chain
   (D) pioneer organisms compete with consumers
   (E) all of the above are correct

14. Here is a sketch. All of the following processes produce this molecule EXCEPT the

(A) Calvin cycle
(B) Krebs cycle
(C) electron transport chain
(D) light-dependent reactions
(E) glycolysis

15. A black animal is crossed with a white animal and all the offspring are black. Which pattern of inheritance is at work?

(A) Law of dominance
(B) Law of segregation
(C) Incomplete dominance
(D) Codominance
(E) Sex-linked inheritance

16. According to the best scientific evidence about evolution, species descended from a common ancestor

(A) slowly and gradually by the accumulation of many small changes
(B) rapidly through divergent evolution alone
(C) rapidly through mutation alone
(D) in spurts of relatively rapid changes
(E) because they needed to adapt to a changing environment or they would die

17. A black hen is crossed with a white rooster, and only gray offspring result. If two of these gray offspring are mated, what is the chance of hatching a white offspring?

(A) 0%
(B) 25%
(C) 50%
(D) 75%
(E) 100%

18. Farmers have successfully bred Brussels sprouts, broccoli, kale, and cauliflower from the mustard plant. This demonstrates

    (A) convergent evolution
    (B) coevolution
    (C) adaptive radiation
    (D) natural selection
    (E) artificial selection

Directions: Each set of lettered choices below refers to the numbered questions or statements immediately following it. Select the one lettered choice that best answers each question and fill in the corresponding space on the answer sheet. A choice may be used once, more than once, or not at all in each set.

Questions 19–21

    (A) Imprinting
    (B) Classical conditioning
    (C) Fixed action pattern
    (D) Altruism
    (E) Operant conditioning

19. Geese hatchlings follow the first thing they see

20. Innate, highly stereotypical behavior, which, once begun, is continued to completion no matter how useless

21. Trial and error learning

22. Scientists believe that the giraffe originally had a short neck that has grown longer over time. The most likely explanation of this is which of the following?

    I. Natural selection
    II. Adaptive radiation
    III. Divergent evolution

    (A) I only
    (B) I and II only
    (C) I and III only
    (D) II and III only
    (E) I, II, and III

23. Which of the following are most closely related?

     I. *Acer rubrum*
    II. *Acer sucre*
   III. *Pseudotriton rubrum*

    (A) I and II
    (B) II and III
    (C) I and III
    (D) All are closely related

Questions 24–26

Refer to this drawing of the eye.

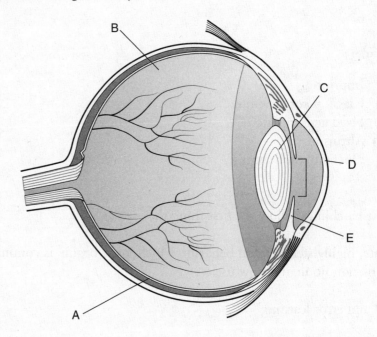

24. Identify the structure that changes to allow different amounts of light to enter the eye

25. Identify the structure that absorbs light and sends nerve impulses to the brain

26. Identify the retina

<u>Questions 27–29</u>

Five beakers are used in an experiment about osmosis. Each beaker contains
50 mL of a sucrose solution of varying concentrations: 0.2 M, 0.4 M, 0.6 M,
0.8 M, or 1.0 M. Pieces of fresh potato (each 10.0 g in mass) are cut up, weighed,
and placed into the beakers. After 12 hours, the potatoes are carefully removed
from each beaker and weighed again. See the data in the table below.

Beaker	Concentration of Sucrose Solution	Mass of Potato at Time Zero	Mass of Potato After 12 Hours
1	0.2 M	10.0 g	8.2 g
2	0.4 M	10.0 g	9.4 g
3	0.6 M	10.0 g	10.8 g
4	0.8 M	10.0 g	11.5 g
5	1.0 M	10.0 g	13.6 g

27. In this experiment

(A) water flowed into the potato only
(B) water flowed out of the potato only
(C) sucrose flowed into the potato only
(D) sucrose flowed both into and out of the potato
(E) water flowed both into and out of the potato

28. Given the results of this experiment, what is the molarity (concentration)
within the potato cells?

(A) Less than 0.2 M
(B) Less than 0.4 M but greater than 0.2 M
(C) Less than 0.6 M but greater than 0.4 M
(D) Less than 0.8 M but greater than 0.6 M
(E) Greater than 0.8 M

29. The results of this experiment give support to the theory that

(A) water diffuses down a gradient
(B) water can be actively transported against a gradient
(C) solutes will diffuse from high concentration to low concentration
(D) living cells respond in different ways to the same conditions
(E) potato cells respond differently from other living cells

Questions 30–32

Refer to the pedigree below that shows inheritance of blood types. (Males are squares; females are circles.)

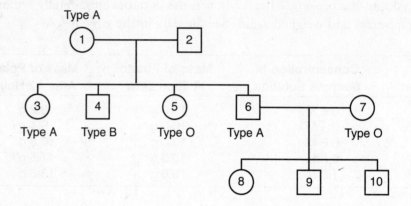

Type A

Type A   Type B   Type O   Type A   Type O

30. What is correct about the inheritance of blood types?

(A) A is dominant over B
(B) B is dominant over A
(C) O is dominant over A only
(D) O is dominant over both A and B
(E) O is recessive

31. The genotype of person #2 is

(A) AO
(B) AA
(C) BO
(D) BB
(E) AB

32. If couple #6 and #7 had another child, what is the chance the blood type would be O?

(A) 0%
(B) 25%
(C) 50%
(D) 75%
(E) 100%

Questions 33–35

Refer to this diagram of the human heart.

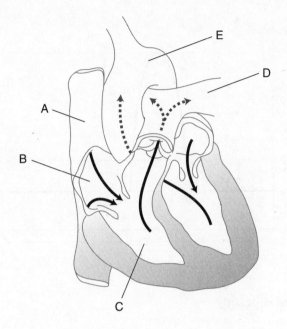

33. The vessel that sends blood to the entire body

34. The chamber that receives blood from the body

35. The artery that carries deoxygenated blood

Questions 36–39

Refer to this graph of an impulse passing across a neuron.

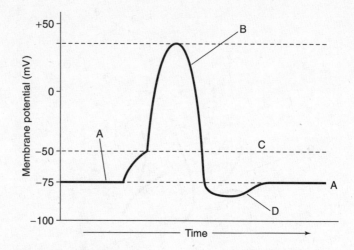

36. The impulse is passing

37. The sodium-potassium pump is responsible for pumping ions across the membrane

38. A steep gradient of sodium and potassium ions exists at the axon membrane

39. An impulse cannot pass

Questions 40–43

A study of a small farm in Michigan was carried out in 2004. A variety of organisms were found to live there, including meadow voles, grasshoppers, spiders, birds, and mice. The farmer retired and moved away, leaving the land to grow wild.

40. The meadow voles, grasshoppers, spiders, mice, and other organisms, along with the soil, minerals, and water make up a(n)

   (A) ecosystem
   (B) population
   (C) community
   (D) food chain
   (E) desert biome

41. The study of the farm revealed the population size of the different species of animals during the summer months of June, July, and August. The results are recorded in the table below.

Species of Animals	Numbers of Organisms		
	June	July	August
Spiders	850	300	550
Grasshoppers	1,800	4,600	4,000
Mice	275	225	250
Birds	95	80	90

Which is correct about the data collected from June through August?

(A) Only the spider population changed to any extent.
(B) The population of mice increased as the summer went on.
(C) The population of grasshoppers remained fairly constant.
(D) The population of birds remained fairly constant.
(E) Both the population of spiders and mice remained constant.

42. What will most likely occur if the farm is sold and the fields are allowed to grow wild?

(A) The plants will change, but the animals will stay the same.
(B) The animals will change, but the plants will stay the same.
(C) Neither the plants nor the animals will change because the climate will not change.
(D) Both the animals and plants will change.
(E) All the animals will slowly die out because they will not be adapted to the new environment.

43. Although several different species of birds inhabit the farm, competition between these birds rarely occurs. The best explanation for this lack of competition is that these birds

(A) share food with each other
(B) have a limited supply of food
(C) live in different ecological niches
(D) are closely related
(E) have experienced mutations in their DNA that prevent them from competing

44. Humans eat corn and other vegetables. Humans also eat beef from cattle that were corn fed. In those cases, cattle and humans occupy which of the following trophic levels?

    (A) Producer and primary consumer
    (B) Primary consumer and secondary consumer
    (C) Secondary consumer and tertiary consumer
    (D) Tertairy consumer and quaternary consumer
    (E) Both are primary consumers

Questions 45–47

    (A) Tundra
    (B) Marine biome
    (C) Desert
    (D) Temperate deciduous forest
    (E) Tropical rain forest

45. Only 4 percent of the land surface but accounts for 20 percent of Earth's food production

46. Provides most of Earth's food and oxygen

47. Consists of trees that drop their leaves in winter

Questions 48–51

Refer to this diagram of a flower.

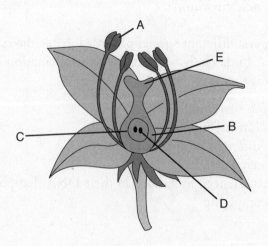

48. Site where the pollen germinates

49. Site of sperm production

50. Becomes the fruit

51. Becomes the seed

Questions 52–54

The table shows a series of four fruit fly experiments breeding normal and vestigial wing flies.

	Parents		Offspring	
Cross	Female	Male	Normal Wing	Vestigial Wing
1	Normal wing	Vestigial wing	95	0
2	Vestigial wing	Normal wing	105	0
3	Normal wing	Normal wing	76	23
4	Normal wing	Vestigial wing	56	51

52. The trait for vestigial wings is most likely

    (A) Autosomal dominant
    (B) Autosomal recessive
    (C) Sex-linked dominant
    (D) Sex-linked recessive
    (E) It cannot be determined

53. What is the most likely genotype of the female normal wing in cross 4?

    (A) *Nn*
    (B) *NN*
    (C) *nn*
    (D) X–X
    (E) X–X–

54. What is the most likely genotype of the male normal wing in cross 3?

(A) *NN*
(B) *Nn*
(C) *nn*
(D) X–X
(E) X–X–

55. Which is true about the karyotype below?

(A) It shows a normal female.
(B) It shows a normal male.
(C) It shows a person who suffers from Down syndrome.
(D) It shows a person who suffers from a gene mutation.
(E) It shows a person who suffers from a condition that results from nondisjunction.

Questions 56–58

Refer to this sketch of prokaryotic DNA as it commonly undergoes replication and transcription simultaneously.

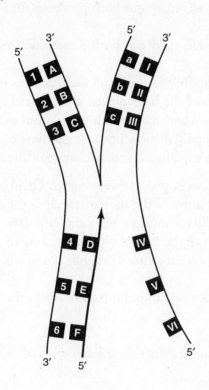

56. If 1 is thymine, then A must be

    (A) guanine
    (B) cytosine
    (C) thymine
    (D) adenine
    (E) uracil

57. If 4 is adenine, then D must be

    (A) guanine
    (B) cytosine
    (C) thymine
    (D) adenine
    (E) uracil

58. In what way would the process shown be different in a eukaryotic cell?

    (A) Eukaryotic cells do not carry out transcription.
    (B) Eukaryotic cells do not carry out replication.
    (C) Eukaryotic cells do not carry out both transcription and replication.
    (D) Eukaryotic cells carry out both processes, but they do not occur at the same time.
    (E) Eukaryotic cells carry out both these processes in the Golgi body.

59. A fungus infection affected nearly all the oak trees in a particular forest so that the coloration of the bark turned almost black. Scientists studying the diseased trees discovered that a moth population that inhabited the forest changed from being light brown to being almost black. Which of the following would best explain that color change of the moth population?

    (A) The moths developed darker wings to blend in with the trees.
    (B) The fungus infected the moths as well as the oak trees.
    (C) The almost black moths within the population were the only ones to survive once the trees darkened because of the fungus infection.
    (D) The moths were the first to change color, which caused the trees to darken.
    (E) The fungus caused mutations to occur in the moths as well as in the oak trees.

60. According to scientific evidence, the age of Earth is closest to

    (A) 400 years old
    (B) 4,000 years old
    (C) 400,000 years old
    (D) 4,000,000 years old
    (E) 4 billion years old

If you are taking the Biology-E test, continue with questions 61–80.
If you are taking the Biology-M test, go to question 81 now.

Biology-E Section

If you are taking the Biology-E test, continue with questions 61–80. Be sure to begin this section of the test by filling in circle 61 on your answer sheet.

61. All of the following are true of organisms classified in the domain Archaea EXCEPT

    (A) one example is *E. coli*, the organism that lives in the human gut
    (B) they can thrive in environments with very high temperatures
    (C) they can thrive in environments with high salt concentrations
    (D) they have no internal membranes
    (E) their DNA can contain introns

62. Factors that influence population density include which of the following?

    I. Predation
    II. Interspecies competition
    III. Intraspecies competition

    (A) I only
    (B) II only
    (C) I and III only
    (D) II and III only
    (E) I, II, and III

63. The human population today can best be described as

    (A) declining
    (B) growing linearly
    (C) growing exponentially
    (D) at the carrying capacity
    (E) fluctuating seasonally

64. All of the following are true of K-strategists EXCEPT

    (A) intensive parenting
    (B) reproduce only once or twice
    (C) example: humans
    (D) large young
    (E) slow maturation

65. Lamprey eels attach to the skin of certain trout and absorb nutrients from the body of the trout. Which symbols best represent this relationship?

    (A) (+/+)
    (B) (+/0)
    (C) (+/−)
    (D) (−/+)
    (E) (−/0)

<u>**Questions 66–67**</u>

When two species of paramecium, *P. caudatum* and *P. aurelia*, are grown in separate culture dishes, each population grows rapidly and then levels off at the carrying capacity for its environment. When they are combined in one culture dish, *P. aurelia* survives, but *P. caudatum* does not.

66. This is most likely because

    (A) *P. caudatum* suffers a mutation that prevents its survival
    (B) *P. caudatum* was attacked by a pathogenic virus or bacteria
    (C) *P. aurelia* must have evolved into a superior organism
    (D) The two populations are competing with each other, and *P. aurelia* can outcompete *P. caudatum*
    (E) *P. aurelia* is a predator, and *P. caudatum* is its prey

67. Which of the following statements is true about the two populations of paramecium grown in culture together?

    (A) They share a niche.
    (B) This is an example of commensalism.
    (C) The two populations do not interact.
    (D) This is an example of divergent evolution.
    (E) This is an example of primary ecological succession.

Questions 68–69

This graph shows the effects of large-scale deforestation on rainfall in the Amazon rain forest.

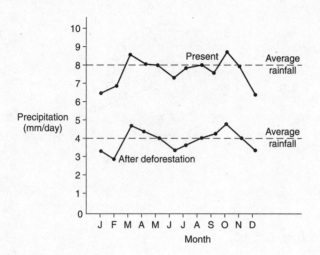

68. Deforestation will ultimately

    (A) cause precipitation to decrease about 25%
    (B) cause precipitation to decrease about 50%
    (C) cause precipitation to increase about 25%
    (D) cause precipitation to increase about 50%
    (E) cause precipitation to increase at first and then decrease in future years

69. What is the most likely explanation for the change in precipitation as a result of the deforestation of a vast area like the Amazon basin?

    (A) Runoff increases and greatly reduces transpiration and the formation of clouds.
    (B) Local temperatures rise due to the reduction of the cooling process of transpiration.
    (C) The length of the rainy season will increase.
    (D) The climate in other parts of the world will be altered.
    (E) Although they know it occurs, scientists do not understand this phenomenon.

<u>**Questions 70–72**</u>

Refer to this sketch that shows early embryonic development.

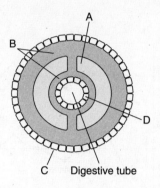

70. The structure at *C* is

    (A) endoderm and will give rise to the nervous system
    (B) mesoderm and will give rise to the coelom
    (C) ectoderm and will give rise to the nervous system
    (D) ectoderm and will give rise to the gut
    (E) endoderm and will give rise to the gut

71. The function of the structure at *A* is to

    (A) give rise to blood and bones
    (B) provide a cavity for organ systems
    (C) give rise to muscle
    (D) give rise to the viscera
    (E) provide an end stage to cleavage

72. This embryo could NOT develop into a(n)

    (A) sea star
    (B) flatworm
    (C) lobster
    (D) snake
    (E) earthworm

Questions 73–74

Nitrogen fertilizer has been applied yearly since 1850 to an experimental farm in the United States. These three graphs show the data collected over 100 years until 1950.

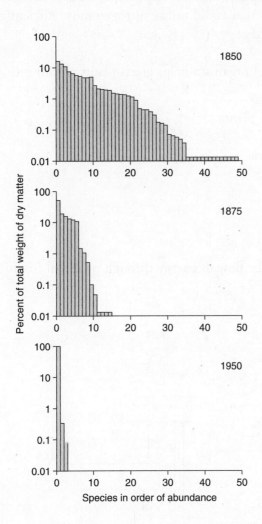

73. Which statement most accurately describes the information shown on these graphs?

   (A) There were the fewest number of species of plants in 1850.
   (B) There was the least biomass in 1850.
   (C) There was a decline in diversity over 100 years.
   (D) In 1875, there were more species of plant but less biomass than in 1850.
   (E) The number of species of plants and the biomass increased from 1850 to 1950.

74. What is the most likely cause of the change over 100 years?

   (A) There was less air pollution and therefore more sunlight in 1850.
   (B) Global warming occurred.
   (C) Natural ecological succession occurred.
   (D) Global warming was responsible for a decrease in biomass.
   (E) Plants that could utilize nitrogen most efficiently had the selective advantage.

75. What type of organism helps green plants absorb nitrogen from the soil?

   (A) Lichens
   (B) Other plants
   (C) Fungi
   (D) Bacteria
   (E) Viruses

Questions 76–78

This sketch shows the flow of energy through principal trophic levels in an ecosystem.

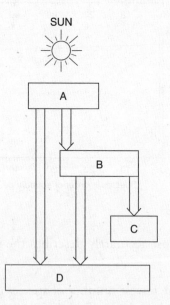

76. Which level has the greatest biomass?

   (A) A
   (B) B
   (C) C
   (D) D

77. Which level has the fewest numbers of organisms?

    (A) A
    (B) B
    (C) C
    (D) D

78. Which organisms most likely occupy level D?

    (A) Bacteria
    (B) Snakes
    (C) Bats
    (D) Cows
    (E) Humans

Questions 79–80

The total amount of energy converted into products of photosynthesis is known as gross primary productivity. Much of this energy is used by the plant to maintain itself. The remaining energy is called net primary productivity, and it provides all the energy for all the food chains in the world.

79. Which factor contributes LEAST to net productivity?

    (A) Light
    (B) Respiration rates
    (C) Temperature
    (D) Rainfall
    (E) Transpiration rates

80. Which terrestrial region would have the highest net productivity?

    (A) Desert
    (B) Tropical rain forest
    (C) Taiga
    (D) Tundra
    (E)   Sandy beach

STOP

If you finish before time is called, you may check your work on the entire biology test only. Do not turn to any other test in the book.

Biology-M Section

If you are taking the Biology-M test, continue with questions 81–100. Be sure to begin this section of the test by filling in circle 81 on your answer sheet.

81. Convert 65.0 nm to mm.

    (A) 650.
    (B) 6,500.
    (C) 65,000.
    (D) 0.065
    (E) 0.65

82. Glucose + fructose → _____

    (A) maltose
    (B) lactose
    (C) sucrose
    (D) galactose
    (E) dextrose

83. Which of the following levels of protein structure directly determines how a particular enzyme will function?

    (A) Primary
    (B) Secondary
    (C) Tertiary
    (D) Quaternary

84. Here is a sketch of a plant cell placed into a solution. Which of the following is correct?

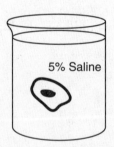

5% Saline

    (A) The solution is hypotonic to the cell, and the cell will swell and burst.
    (B) The solution is hypotonic to the cell, and the cell will become turgid.
    (C) The solution is hypertonic to the cell, but nothing will happen to the cell.
    (D) The solution is hypertonic to the cell, and the cell will shrink.
    (E) The solution is hypertonic to the cell, and the cell will swell.

85. If a solution has a pH of 2, then it has _____ moles of H^+ ions per liter.

   (A) 0.2
   (B) 0.02
   (C) 2.0
   (D) 0.1
   (E) 0.01

86. Which of the following does NOT require ATP?

   (A) Facilitated diffusion
   (B) Activation of the contractile vacuole
   (C) Sodium-potassium pump
   (D) Receptor-mediated endocytosis
   (E) Pinocytosis

87. All of the following are correct about microscopes EXCEPT

   (A) resolving power is the same thing as magnification
   (B) total magnification is determined by multiplying the magnification of the ocular lens by the magnification of the objective lens
   (C) an electron microscope requires a vacuum column
   (D) living tissue cannot generally be studied in an electron microscope
   (E) the scanning electron microscope is best used to study the surfaces of cells

88. This molecule would NOT be a component of

   (A) maltase
   (B) hemoglobin
   (C) glycogen
   (D) insulin
   (E) a plasma membrane

89. All of the following are correct about telomeres EXCEPT

   (A) they are created by the enzyme telomerase
   (B) they protect the ends of chromosomes
   (C) the more divisions a cell undergoes, the shorter the telomeres
   (D) they consist of nonsense nucleotide sequences
   (E) they assist in the synthesis of proteins

90. All of the following are correct about RNA processing EXCEPT

    (A) it occurs in the nucleus
    (B) it occurs after transcription
    (C) it occurs before translation
    (D) introns are added
    (E) noncoding regions are removed from the RNA

Questions 91–93

Refer to the terms below.

    (A) Thymine
    (B) Deoxyribose
    (C) Ribose
    (D) Uracil
    (E) Guanine

91. Sugar found in RNA but not in DNA

92. Nitrogneous base found in RNA but not in DNA

93. Nitrogenous base that occurs with the same frequency as cytosine

94. All of the following statements about gel electrophoresis are correct EXCEPT:

    (A) restriction enzymes cut DNA at particular recognition sequences
    (B) restriction enzymes are extracted from bacteria
    (C) the farther a piece of DNA runs, the larger it is
    (D) *EcoRI* is an example of a restriction enzyme
    (E) restriction enzymes are necessary in order to insert a gene into a plasmid

95. Here is a fragment of DNA that has been cut at certain points and will be run through a gel. In which order will the DNA fragments run, beginning at the wells?

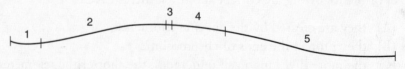

Fragment of DNA

    (A) 1, 2, 3, 4, 5
    (B) 5, 4, 3, 2, 1
    (C) 5, 2, 4, 1, 3
    (D) 5, 2, 3, 1, 4
    (E) 3, 1, 4, 2, 5

96. Here is a sketch of the field of view under a light microscope at 100x magnification. The diameter of the field is 400 nm. What is the approximate length of the cell in the field?

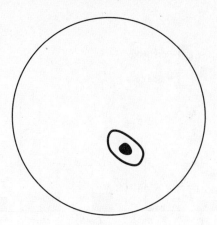

   (A) 20 nm
   (B) 40 nm
   (C) 60 nm
   (D) 80 nm
   (E) 100 nm

97. Two brothers were under medical treatment for infertility. Microscopic examination of their semen shows that although the sperm looked normal, they did not move properly. The brothers also suffered with chronic bronchitis. The doctors studying the cases decided that both men had a problem with one particular cell organelle. Which one?

   (A) Endoplasmic reticulum
   (B) Golgi body
   (C) Ribosomes
   (D) Microtubules
   (E) Mitochondria

98. Given the concept that cells are modified for their particular function, which cell organelle would be unusually plentiful in white blood cells?

   (A) Lysosomes
   (B) Golgi bodies
   (C) Ribosomes
   (D) Mitochondria
   (E) Endoplasmic reticulum

99. This bar graph shows the relative concentrations of different ions in pond water and in the cytoplasm of the algae *Chlamydomonous*. Which statement best describes what this graph indicates?

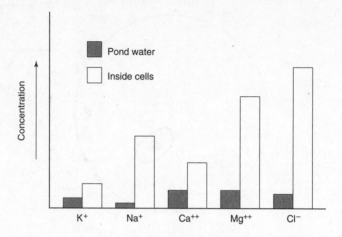

(A) *Chlamydomonous* belongs to the kingdom Protista.
(B) The pond water is hypertonic to the *Chlamydomonous* cells.
(C) Only Mg^+ and Cl^- are absorbed by active transport.
(D) The *Chlamydomonous* has contractile vacuoles that pump out excess water.
(E) *Chlamydomonous* cells can regulate what materials they exchange with their surroundings.

100. Which cell organelle in *Chlamydomonous* is LEAST involved in transport in single-celled organisms living in pond water?

(A) Mitochondria
(B) Chloroplast
(C) Plasma membrane
(D) Contractile vacuole
(E) Nucleus

STOP

Determine Your Raw Score

Step 1: Compare your answer sheet to the correct answers on Table 20.1.

- Put a check in the column marked "Right" if your answer is correct.
- Put a check in the column marked "Wrong" if your answer is incorrect.
- Leave both columns blank if you omitted the question.

Step 2: Count the number of right answers, and enter the number here: _____

Step 3: Count the number of wrong answers, divide by 4, and enter the number here: _____

Step 4: Subtract the number you obtained in Step 3 from the number in Step 2. Round the result to the nearest whole number (0.5 is rounded up), and enter here: _____

Step 5: The number you obtained in Step 4 is your raw score. Convert it to your College Board Score using Tables 20.2 and 20.3, which are similar to those published by The College Board.

TABLE 20.1

The Correct Answers

Question No.	Correct Answer	Right	Wrong	Question No.	Correct Answer	Right	Wrong
1	D			26	A		
2	C			27	E		
3	B			28	C		
4	E			29	A		
5	D			30	E		
6	C			31	C		
7	A			32	C		
8	A			33	E		
9	D			34	B		
10	D			35	D		
11	E			36	B		
12	B			37	D		
13	A			38	A		
14	A			39	D		
15	A			40	A		
16	D			41	D		
17	B			42	D		
18	E			43	C		
19	A			44	B		
20	C			45	E		
21	E			46	B		
22	A			47	D		
23	A			48	E		
24	E			49	A		
25	A			50	B		

TABLE 20.1 (continued)

The Correct Answers

Question No.	Correct Answer	Right	Wrong	Question No.	Correct Answer	Right	Wrong
51	C			76	A		
52	B			77	C		
53	A			78	A		
54	B			79	B		
55	E			80	B		
56	D			81	D		
57	E			82	C		
58	D			83	C		
59	C			84	D		
60	E			85	E		
61	A			86	A		
62	E			87	A		
63	C			88	C		
64	B			89	E		
65	C			90	D		
66	D			91	C		
67	A			92	D		
68	B			93	E		
69	A			94	C		
70	C			95	C		
71	B			96	E		
72	B			97	D		
73	C			98	A		
74	E			99	E		
75	D			100	B		

TABLE 20.2

Score Conversion Table Biology-E

Raw Score	Scaled Score	Raw Score	Scaled Score	Raw Score	Scaled Score
80	800	50	630	20	440
79	800	49	620	19	430
78	800	48	610	18	430
77	800	47	600	17	420
76	800	46	590	16	410
75	800	45	590	15	410
74	800	44	580	14	400
73	800	43	580	13	390
72	790	42	570	12	390
71	780	41	570	11	380
70	780	40	560	10	380
69	770	39	560	9	370
68	760	38	550	8	360
67	750	37	550	7	360
66	750	36	540	6	360
65	740	35	540	5	350
64	730	34	530	4	340
63	720	33	530	3	330
62	710	32	520	2	330
61	700	31	510	1	320
60	690	30	500	0	310
59	690	29	490	−1	300
58	680	28	490	−2	290
57	670	27	480	−3	290
56	670	26	470	−4	280
55	660	25	470	−5	270
54	660	24	460	−6	270
53	650	23	460	−7	260
52	650	22	450	−8	250
51	640	21	440	−9	250

TABLE 20.3

Score Conversion Table Biology-M

Raw Score	Scaled Score	Raw Score	Scaled Score	Raw Score	Scaled Score
80	800	50	630	20	440
79	800	49	620	19	430
78	800	48	610	18	430
77	800	47	600	17	420
76	800	46	590	16	410
75	790	45	590	15	410
74	780	44	580	14	400
73	780	43	580	13	390
72	770	42	570	12	390
71	770	41	570	11	380
70	760	40	560	10	380
69	760	39	560	9	370
68	750	38	550	8	360
67	740	37	550	7	360
66	740	36	540	6	360
65	730	35	540	5	350
64	720	34	530	4	340
63	710	33	530	3	330
62	700	32	520	2	330
61	700	31	510	1	320
60	690	30	500	0	310
59	690	29	490	−1	300
58	680	28	490	−2	290
57	670	27	480	−3	290
56	670	26	470	−4	280
55	660	25	470	−5	270
54	660	24	460	−6	270
53	650	23	460	−7	260
52	650	22	450	−8	250
51	640	21	440	−9	250

EXPLANATION OF ANSWERS

1. **(D)** Choices (A), (B), and (E) have nothing to do with the answers. Choice (C) is false. Some eukaryotic cells, like yeast, reproduce asexually.

2. **(C)** Some plants are prevented or isolated from reproducing with other plants. The cause of the isolation is not important.

3. **(B)** Variation in a population results from an influx or development of new genetic material. Conjugation is a primitive form of sexual reproduction carried out by bacteria and algae. If a population is isolated, there can be no flow of genetic material.

4. **(E)** During the light-dependent reactions, water is broken apart, a process called photolysis, in order to provide electrons that were lost from chlorophyll *a* in photosystem II.

5. **(D)** A blue jay is a bird. Although birds are homeotherms (warm-blooded) and also classified as chordates and vertebrates, as mammals are, they are not mammals.

6. **(C)** Density-independent means not related to the size of a population. In general, flooding is a natural occurrence, not the result of interference by humans. All the rest are density-dependent factors that result from overpopulation.

7. **(A)** According to the Hardy-Weinberg theorem, the characteristics of a stable, nonevolving population include no mutation, a large population, no migration into or out, and random mating.

8. **(A)** The characteristics of a plasma membrane differ with the type of cell. The membrane of a neuron or skeletal muscle cell must have different properties from the membrane of a skin cell. Remember, form and function are related.

9. **(D)** Fish and amphibians produce large numbers of eggs where fertilization and development is external and there is no parenting. In reptiles, fertilization is internal and development is external, and usually there is little or no parenting. Birds are similar to reptiles in fertilization and development, but there is usually a great deal of parenting.

10. **(D)** A solution of pH 5 has a H^+ concentration of 1×10^{-5} M or -0.00001 M. A solution of pH 7 has a H^+ concentration of 1×10^{-7} M or -0.0000001 M. You can see that 0.00001 M is 100 times more concentrated than 0.0000001 M.

11. **(E)** Vitamins are coenzymes. Along with minerals, which act as cofactors, they enable enzymes to function properly. We need very small amounts of vitamins and minerals for normal enzyme function.

12. **(B)** Tendons connect bone to muscle; ligaments connect bone to bone.

13. **(A)** About 10% of the energy stored in the organic matter of any trophic level is converted to organic matter at the next tropic level. Therefore the length of food chains is limited to 4 or 5 tropic levels.

14. **(A)** The molecule is ATP. The Calvin cycle is the process by which plants make sugar. It occurs during the light-independent reactions and requires enormous amounts of ATP. All the other processes produce ATP.

15. **(A)** If a black animal is crossed with a white animal and all the offspring are black, that means that black is dominant and white is recessive.

16. **(D)** The current theory which was developed by Eldridge and Gould, is that evolution occurred by a process called punctuated equilibrium. According to this model, species diverge in spurts of relatively rapid changes instead of slowly and gradually as Darwin hypothesized. Lamarck believed that organisms evolved because they needed to. His theory was rejected long ago.

17. **(B)** The pattern of inheritance at work here is incomplete dominance or blending. The gray animals have the genotype BW, where neither trait dominates over the other.

    Here is the cross:

	B	W
B	BB	BW
W	BW	WW

    There is a 25% chance of producing a white flower, *WW*.

18. **(E)** Artificial selection is the selective breeding of domesticated plants and animals to develop desired traits.

19. **(A)** Imprinting is a type of learning that is responsible for the bonding between mother and offspring. It occurs in a sensitive or critical period in early life.

20. **(C)** Fixed action pattern is a stereotypical behavior. An example is the behavior of hens to roll their eggs periodically. This is necessary for the embryo to develop normally; however, the hen will roll anything that looks remotely like an egg.

21. **(E)** Operant conditioning is also called trial and error learning. An animal learns to associate one of its own behaviors with a reward or punishment and then repeats or avoids the behavior.

22. **(A)** Only natural selection explains how the giraffe got its long neck. Ages ago when food was more plentiful, some giraffes had long necks and some had short necks. Over time, giraffes began to compete for limited food. The animals with longer necks did better than the ones with shorter necks. Only long-necked giraffes passed on their genes, and soon all giraffes had longer necks.

23. **(A)** *Acer rubrum* (red maple) and *Acer sucre* (sugar maple) belong to the same genus, along with all maple trees. Choice I and choice III are not related to *Pseudotriton rubrum*, which is a salamander. *Rubrum* merely means red.

24. **(E)** Light enters the eye through the opening called the pupil. However, the iris, the colored part of the eye, controls how much light can enter by increasing or decreasing the size of the opening called the pupil.

25. **(A)** The retina is located in the back of the eyeball. It absorbs light waves and sends the impulses to the brain.

26. **(A)** The retina is located in the back of the eyeball. It absorbs light waves and sends the impulses to the brain.

27. **(E)** Potato samples 1 and 2 gained weight because water diffused into the potato. Samples 3, 4, and 5 lost weight because water diffused out of the cells.

28. (C) The molarity must be somewhere between that of samples 2 and 3 because that would represent the point of equilibrium.

29. (A) There was no evidence that anything but water diffused into or out of the potato cells. Water always diffuses from high concentration to low concentration.

30. (E) Both types A and B are dominant over O, and they are codominant with respect to each other.

31. (C) Since person #2 and person #1 have children with blood types O (*ii*, the recessive genotype) as well as type B, person #2 must have type BO blood.

32. (C) Parent #6 is the son of #1 and #2 who were AO and BO, respectively. So #6 is AO, having inherited blood gene A from his mom (#1) and gene O from his dad (#2). Parent #7 must be *ii*, so the chance of having a child with blood type O is 50%.

33. (E) The aorta sends blood to the entire body.

34. (B) The chamber that receives blood from the entire body is the right atrium.

35. (D) The pulmonary artery carries deoxygenated blood from the right ventricle to the lungs.

36. (B) The impulse is passing at what is known as the action potential. The membrane becomes depolarized.

37. (D) The refractory period is the period of time when the sodium-potassium pump returns the membrane to its resting potential. No impulse can pass at that time.

38. (A) When the membrane is at rest (resting potential), the membrane is polarized with sodium and potassium on opposite sides of the membrane.

39. (D) The refractory period is the time when the sodium-potassium pump returns the membrane to its resting potential. No impulse can pass at that time.

40. (A) An ecosystem consists of all the living things and the nonliving things they interact with in one environment.

41. (D) By looking at the chart, only one species remained relatively constant in terms of population: birds.

42. (D) Since the environment will change, both the plants and animals will change. The plants will change in response to the change in environment, and the animals will change in response to the change in plants.

43. (C) Animals that share a niche, meaning they eat the same food, compete with each other. Since these birds all live in the same area but do not compete, they must not be sharing a niche.

44. (B) Animals can occupy different trophic levels in different food chains. When cattle eat corn, they are acting as primary consumers. When humans eat beef, we are acting as secondary consumers. However, we can also eat corn and also act as primary consumers.

45. (E) The tropical rain forest produces the most food of any terrestrial biome. The marine biome produces the most food if you include all the biomes.

46. (B) The marine biome produces the most food of all biomes. Because it contains a vast amount of producers, it produces the most oxygen of any biome.

47. **(D)** The temperate deciduous forest is also known as the boreal forest.

48. **(E)** The stigma, the top of the pistil, is sticky and is the site where the pollen lands and germinates.

49. **(A)** Each pollen grain contains 3 monoploid nuclei. The production of these gametes occurs in the anthers, part of the stamen.

50. **(B)** The ripened ovary (after fertilization) becomes the fruit.

51. **(C)** After fertilization, the ovule becomes the seed. The seed contains the embryo and food for the growing embryo.

52. **(B)** Crosses 1 and 2 between a normal and vestigial wing always produce normal wing flies. Whether the male or female parent has vestigial wings does not affect the outcome. This means that the trait is autosomal. Since all the F_1 have normal wings, vestigial must be recessive.

53. **(A)**

	N	*n*
n	Nn	nn
n	Nn	nn

54. **(B)**

	N	*n*
N	NN	Nn
n	Nn	nn

55. **(E)** There are only 45 chromosomes and the Y chromosome is missing from this pedigree. This must have occurred as a result of nondisjunction; one gamete received 22 chromosomes, while the other gamete received 24 chromsomes.

56. **(D)** The top of the picture shows replication because there are two old strands each forming two new strands. The bottom part of the picture shows transcription. You know this because only one strand of DNA is operating. Thymine always pairs with adenine in DNA.

57. **(E)** This is the area where DNA makes RNA. So the strand DEF is RNA. If 4 is adenine, then D is uracil because there is no thymine in RNA.

58. **(D)** In prokaryotes, replication occurs at the same time as transcription. There is no nuclear membrane to separate where the two processes occur. In eukaryotes, replication occurs in the nucleus and transcription occurs in the cytoplasm. Additionally, the two processes occur at different times.

59. **(C)** Prior to the infection by fungi, oak trees were brown and light brown moths were camouflaged. When the trees darkened due to fungus infection, the moth population darkened, as light brown moths were eaten because they could be seen by predators. This is the process of natural selection. There is no evidence that the fungus infection cause a mutation.

60. **(E)** Using the best scientific information we have, the Earth is calculated to be 4.6 billion years old.

61. **(A)** The bacteria *E. coli* belongs in the domain Bacteria, not Archaea. All the other statements are true about the domain Archaea.

62. **(E)** All of these factors influence population density.

63. **(C)** The human population is growing exponentially.

64. **(B)** Humans are K-strategists, and we can reproduce many times. R-strategists like insects only reproduce once, although they can lay hundreds or thousands of eggs at that time. Their lives are short, and they do not have the chance to reproduce again.

65. **(C)** The lamprey eel benefits, and the fish suffers. The eel is a parasite.

66. **(D)** When two organisms of the same species compete for food or other resources, it is because they share a niche. The more fit one will survive at the expense of the other one. This is known as Gause's principle of exclusion.

67. **(A)** When two organisms of the same species compete for food or other resources, it is because they share a niche. The more fit one will survive at the expense of the other one. This is known as Gause's principle of exclusion.

68. **(B)** The average rainfall prior to deforestation was 8 mm per day. After deforestation, it is 4 mm per day, on average. That is a 50% decrease.

69. **(A)** Trees normally give off large quantities of water by transpiration, which turns into clouds, which turns into rain. This is the reason that tropical rain forests are so damp. If there are no trees, there is no transpiration, no cloud formation, and no rain.

70. **(C)** The structure at *C* is the ectoderm, which gives rise to skin and the nervous system.

71. **(B)** The structure at *A* is the coelom, which provides space for organ systems.

72. **(B)** Flatworms, like planaria, do not have a coelom, and they are called acoelomates. Their bodies are flat, and they have simplified body systems that fit between 3 layers of tissue: endoderm, mesoderm, and ectoderm.

73. **(C)** The graph shows only abundance of species. See the label on the *x*-axis.

74. **(E)** The change is the number of different species shown in the graph. Although you may think that modern-day problems like global warming and air pollution are responsible for any change in the environment, ecology is more complex than that. What occurred was ecological succession in reverse. In this case, plants that could utilize nitrogen most effectively were selected, and most species died out.

75. **(D)** Bacteria that live in the nodules of the roots of some plants assist in the uptake of nitrogen. Nitrogen is necessary for plants to make protein.

76. **(A)** Block A represents the producers absorbing sunlight and changing it into sugar. Producers have the most biomass in trophic levels.

77. **(C)** In this case, B represents primary consumers, and C represents secondary consumers. There are fewer secondary consumers than primary consumers.

78. **(A)** Layer D represents bacteria and fungi that cycle all the nutrients in the ecosystem.

79. **(B)** Net productivity is calculated as the gross productivity minus respiration rates. Light, temperature, and rainfall all contribute to the production of food, which is measured by gross productivity, the total rate of food production. Transpiration rates are a measure, in part, of photosynthesis rates.

80. **(B)** Tropical rain forests are the most productive terrestrial biomes. Deserts are the least productive.

81. **(D)** 1 mm = 1,000 nm or 1 nm = 0.001 mm

82. **(C)** Glucose + glucose → maltose; glucose + galactose → lactose

    Glucose, galactose, and fructose are monosaccharides. Sucrose, maltose, and lactose are disaccharides. Dextrose is the same thing as glucose.

83. **(C)** Tertiary structure is superimposed on the patterns of secondary structure. It results from the interactions between the amino acids within the molecule. It gives every protein its particular shape, which determines its unique function.

84. **(D)** The cell is in a 5% solution that is hypertonic to any cell. Water will flow out of the cell into the surrounding solution, and the cell will shrink.

85. **(E)** The pH is the negative log of the hydrogen ion concentration. So pH 2 means 1×10^{-2} M, which equals 0.01 M or 0.01 moles per liter.

86. **(A)** Facilitated diffusion is still diffusion, and diffusion does not require the input of any energy or ATP. All the other choices require input of ATP.

87. **(A)** Resolving power, or resolution, is a measure of the clarity of the image and is not the same thing as magnification. Resolution is the ability to see two dots that are close together as separate. The other choices are all correct statements about microscopes.

88. **(C)** This molecule is an amino acid. Glycogen is a polysaccharide and consists of many glucose molecules (or monosaccharides) strung together. The other choices are all proteins and therefore consist of chains of amino acids.

89. **(E)** All statements about telomeres are correct except (E).

90. **(D)** Introns, the noncoding intervening sequences, are removed from the mRNA strand before it leaves the nucleus. SnRNPs and splicesomes assist in this process. Only the exons (the expressed sequences or genes) leave the nucleus.

91. **(C)** The sugar in RNA is ribose. The sugar in DNA is deoxyribose.

92. **(D)** Uracil replaces thymine in RNA.

93. **(E)** Since guanine and cytosine always pair up together, the number of one must equal the number of the other.

94. **(C)** The farther a piece of DNA runs in a gel, the smaller the piece of DNA it is.

95. **(C)** When DNA that has been cut with restriction enzymes runs through a gel, the shorter pieces run farther and the larger pieces run a shorter distance. Piece 5 is the longest, then piece 2, then 4, then 1, and finally piece 3.

96. **(E)** Use the edge of a piece of paper (since you will not be allowed to have a ruler on the test day) to estimate how many times a length of the cell will go across the diameter of the field. It seems to be about 4 times. If the diameter of the field is 400 nm, then divide the diameter by 4. The answer is closest to 100.

97. **(D)** It is stated that both men have sperm that do not move properly and chronic bronchitis. Microtubules make up cilia and flagella in a 9 plus 2 configuration. Lungs are lined with cilia whose job is to rid the respiratory system

of foreign bodies. There is probably something wrong with their microtubules, which would cause a problem with both the sperm and the lining of the lungs.

98. **(A)** The job of lysosomes is to hydrolyze and destroy particles in the cell. The job of white blood cells is to engulf invaders. Therefore, a prominent organelle in white blood cells is lysosomes, which would aid in the destruction of invaders.

99. **(E)** The bar graph indicates that the concentration inside the cells is different from the concentration outside the cells. This is because the cells can regulate what they exchange with their surroundings.

100. **(B)** Transport is the process by which substances are moved into and out of cells. Mitochondria, plasma membranes, and contractile vacuoles are all directly involved with active transport, which requires energy. The nucleus ultimately controls what goes on in a cell. If the nucleus of a cell is destroyed while leaving the cell intact, active transport ceases. Although chloroplasts make sugar, they do not produce ATP and are the least involved with active transport.

What Topics Do You Need to Work On?

Table 20.4 shows an analysis by topic for each question on the test you just took.

TABLE 20.4

Topic Analysis

Cellular and Molecular Biology	Heredity	Evolution and Diversity	Organismal Biology	Ecology
8, 10, 11, 14, 27, 28, 29, 33, 56, 57, 58, 81–100	1, 15, 17, 22, 30, 31, 32, 52, 53, 54, 55	2, 3, 5, 7, 9, 16, 18, 22, 23, 59, 60	4, 12, 19, 20, 21, 24, 25, 26, 34, 35, 36, 37, 38, 39, 48, 49, 50, 51	6, 13, 40, 41, 42, 43, 44, 45, 46, 47, 61–80

Answer Sheet 2
SAMPLE TEST 2

1 Ⓐ Ⓑ Ⓒ Ⓓ Ⓔ   16 Ⓐ Ⓑ Ⓒ Ⓓ Ⓔ   31 Ⓐ Ⓑ Ⓒ Ⓓ Ⓔ   46 Ⓐ Ⓑ Ⓒ Ⓓ Ⓔ
2 Ⓐ Ⓑ Ⓒ Ⓓ Ⓔ   17 Ⓐ Ⓑ Ⓒ Ⓓ Ⓔ   32 Ⓐ Ⓑ Ⓒ Ⓓ Ⓔ   47 Ⓐ Ⓑ Ⓒ Ⓓ Ⓔ
3 Ⓐ Ⓑ Ⓒ Ⓓ Ⓔ   18 Ⓐ Ⓑ Ⓒ Ⓓ Ⓔ   33 Ⓐ Ⓑ Ⓒ Ⓓ Ⓔ   48 Ⓐ Ⓑ Ⓒ Ⓓ Ⓔ
4 Ⓐ Ⓑ Ⓒ Ⓓ Ⓔ   19 Ⓐ Ⓑ Ⓒ Ⓓ Ⓔ   34 Ⓐ Ⓑ Ⓒ Ⓓ Ⓔ   49 Ⓐ Ⓑ Ⓒ Ⓓ Ⓔ
5 Ⓐ Ⓑ Ⓒ Ⓓ Ⓔ   20 Ⓐ Ⓑ Ⓒ Ⓓ Ⓔ   35 Ⓐ Ⓑ Ⓒ Ⓓ Ⓔ   50 Ⓐ Ⓑ Ⓒ Ⓓ Ⓔ
6 Ⓐ Ⓑ Ⓒ Ⓓ Ⓔ   21 Ⓐ Ⓑ Ⓒ Ⓓ Ⓔ   36 Ⓐ Ⓑ Ⓒ Ⓓ Ⓔ   51 Ⓐ Ⓑ Ⓒ Ⓓ Ⓔ
7 Ⓐ Ⓑ Ⓒ Ⓓ Ⓔ   22 Ⓐ Ⓑ Ⓒ Ⓓ Ⓔ   37 Ⓐ Ⓑ Ⓒ Ⓓ Ⓔ   52 Ⓐ Ⓑ Ⓒ Ⓓ Ⓔ
8 Ⓐ Ⓑ Ⓒ Ⓓ Ⓔ   23 Ⓐ Ⓑ Ⓒ Ⓓ Ⓔ   38 Ⓐ Ⓑ Ⓒ Ⓓ Ⓔ   53 Ⓐ Ⓑ Ⓒ Ⓓ Ⓔ
9 Ⓐ Ⓑ Ⓒ Ⓓ Ⓔ   24 Ⓐ Ⓑ Ⓒ Ⓓ Ⓔ   39 Ⓐ Ⓑ Ⓒ Ⓓ Ⓔ   54 Ⓐ Ⓑ Ⓒ Ⓓ Ⓔ
10 Ⓐ Ⓑ Ⓒ Ⓓ Ⓔ   25 Ⓐ Ⓑ Ⓒ Ⓓ Ⓔ   40 Ⓐ Ⓑ Ⓒ Ⓓ Ⓔ   55 Ⓐ Ⓑ Ⓒ Ⓓ Ⓔ
11 Ⓐ Ⓑ Ⓒ Ⓓ Ⓔ   26 Ⓐ Ⓑ Ⓒ Ⓓ Ⓔ   41 Ⓐ Ⓑ Ⓒ Ⓓ Ⓔ   56 Ⓐ Ⓑ Ⓒ Ⓓ Ⓔ
12 Ⓐ Ⓑ Ⓒ Ⓓ Ⓔ   27 Ⓐ Ⓑ Ⓒ Ⓓ Ⓔ   42 Ⓐ Ⓑ Ⓒ Ⓓ Ⓔ   57 Ⓐ Ⓑ Ⓒ Ⓓ Ⓔ
13 Ⓐ Ⓑ Ⓒ Ⓓ Ⓔ   28 Ⓐ Ⓑ Ⓒ Ⓓ Ⓔ   43 Ⓐ Ⓑ Ⓒ Ⓓ Ⓔ   58 Ⓐ Ⓑ Ⓒ Ⓓ Ⓔ
14 Ⓐ Ⓑ Ⓒ Ⓓ Ⓔ   29 Ⓐ Ⓑ Ⓒ Ⓓ Ⓔ   44 Ⓐ Ⓑ Ⓒ Ⓓ Ⓔ   59 Ⓐ Ⓑ Ⓒ Ⓓ Ⓔ
15 Ⓐ Ⓑ Ⓒ Ⓓ Ⓔ   30 Ⓐ Ⓑ Ⓒ Ⓓ Ⓔ   45 Ⓐ Ⓑ Ⓒ Ⓓ Ⓔ   60 Ⓐ Ⓑ Ⓒ Ⓓ Ⓔ

E Section

61 Ⓐ Ⓑ Ⓒ Ⓓ Ⓔ   66 Ⓐ Ⓑ Ⓒ Ⓓ Ⓔ   71 Ⓐ Ⓑ Ⓒ Ⓓ Ⓔ   76 Ⓐ Ⓑ Ⓒ Ⓓ Ⓔ
62 Ⓐ Ⓑ Ⓒ Ⓓ Ⓔ   67 Ⓐ Ⓑ Ⓒ Ⓓ Ⓔ   72 Ⓐ Ⓑ Ⓒ Ⓓ Ⓔ   77 Ⓐ Ⓑ Ⓒ Ⓓ Ⓔ
63 Ⓐ Ⓑ Ⓒ Ⓓ Ⓔ   68 Ⓐ Ⓑ Ⓒ Ⓓ Ⓔ   73 Ⓐ Ⓑ Ⓒ Ⓓ Ⓔ   78 Ⓐ Ⓑ Ⓒ Ⓓ Ⓔ
64 Ⓐ Ⓑ Ⓒ Ⓓ Ⓔ   69 Ⓐ Ⓑ Ⓒ Ⓓ Ⓔ   74 Ⓐ Ⓑ Ⓒ Ⓓ Ⓔ   79 Ⓐ Ⓑ Ⓒ Ⓓ Ⓔ
65 Ⓐ Ⓑ Ⓒ Ⓓ Ⓔ   70 Ⓐ Ⓑ Ⓒ Ⓓ Ⓔ   75 Ⓐ Ⓑ Ⓒ Ⓓ Ⓔ   80 Ⓐ Ⓑ Ⓒ Ⓓ Ⓔ

M Section

81 Ⓐ Ⓑ Ⓒ Ⓓ Ⓔ   86 Ⓐ Ⓑ Ⓒ Ⓓ Ⓔ   91 Ⓐ Ⓑ Ⓒ Ⓓ Ⓔ   96 Ⓐ Ⓑ Ⓒ Ⓓ Ⓔ
82 Ⓐ Ⓑ Ⓒ Ⓓ Ⓔ   87 Ⓐ Ⓑ Ⓒ Ⓓ Ⓔ   92 Ⓐ Ⓑ Ⓒ Ⓓ Ⓔ   97 Ⓐ Ⓑ Ⓒ Ⓓ Ⓔ
83 Ⓐ Ⓑ Ⓒ Ⓓ Ⓔ   88 Ⓐ Ⓑ Ⓒ Ⓓ Ⓔ   93 Ⓐ Ⓑ Ⓒ Ⓓ Ⓔ   98 Ⓐ Ⓑ Ⓒ Ⓓ Ⓔ
84 Ⓐ Ⓑ Ⓒ Ⓓ Ⓔ   89 Ⓐ Ⓑ Ⓒ Ⓓ Ⓔ   94 Ⓐ Ⓑ Ⓒ Ⓓ Ⓔ   99 Ⓐ Ⓑ Ⓒ Ⓓ Ⓔ
85 Ⓐ Ⓑ Ⓒ Ⓓ Ⓔ   90 Ⓐ Ⓑ Ⓒ Ⓓ Ⓔ   95 Ⓐ Ⓑ Ⓒ Ⓓ Ⓔ   100 Ⓐ Ⓑ Ⓒ Ⓓ Ⓔ

Biology E/M Sample Test 2

With Answers and Analysis

> **Directions:** Each question or incomplete statement is followed by five possible answers or completions. For both Biology-E and Biology-M, select the one choice that is the best answer and fill in the corresponding space on the answer sheet.

1. A college student studying genetics is assigned a semester-long project by her professor, who gives her 4 black and 4 white guinea pigs. There are 2 male and 2 female guinea pigs of each color. The professor tells the student that the guinea pigs in each group are all of the same genotype. The project is to determine the genotype of the black guinea pigs. The student does some preliminary library research and discovers that white fur in guinea pigs is recessive and that black fur is dominant. Which of the following procedures would be the best one for the student to carry out to determine the genotype of the black guinea pigs? (Do not consider cost as a factor in this study.)

    (A) Karyotype
    (B) Blood analysis of the chromosomes to see if there is evidence of nondisjunction
    (C) Analysis with an electron microscope to see if any of the black fur contains any white flecks that cannot be seen with the eye or with a light microscope
    (D) Cross any two of the black animals to see if any offspring are white
    (E) Cross any of the black animals with a white animal to see if any of the offspring are white

2. Fixed action patterns are initiated by external stimuli called

    (A) agonistic behavior
    (B) altruistic behavior
    (C) sign stimuli
    (D) associative learning
    (E) initial visual imprinting

3. Lichens

   (A) are green plants
   (B) are algae
   (C) are the major producer in the open ocean
   (D) are a mutualistic association between algae and fungi
   (E) are included in the domain Archaea

4. All of the following are abiotic factors in an ecosystem EXCEPT

   (A) sunlight
   (B) wind
   (C) soil
   (D) temperature
   (E) producers

5. Bacteria that live in the nodules in the roots of legumes and convert free nitrogen into the ammonium ion are called

   (A) bacteria of decay
   (B) nitrogen-fixing bacteria
   (C) nitrifying bacteria
   (D) denitrifying bacteria
   (E) *E. coli*

6. All of the following are correct about evolution EXCEPT

   (A) geographic isolation can cause the formation of new species
   (B) the best-adapted organisms survive
   (C) if the environment changes suddenly and an organism is not adapted for the new environment, it might die
   (D) new combinations of genes can produce new phenotypes
   (E) single individuals, not populations, can change in response to a changing environment

7. Which of the following conditions results from a chromosome mutation?

   (A) Huntington's disease
   (B) Sickle cell anemia
   (C) Cystic fibrosis
   (D) Down syndrome
   (E) PKU

8. In humans, the rate of breathing is primarily controlled by the

   (A) lungs
   (B) cerebrum
   (C) medulla oblongata
   (D) cerebellum
   (E) pleura

9. A man who is color-blind marries a woman who has normal color vision and does not carry the trait for color blindness. What statement will be true of their children?

   (A) All their sons will be normal, and all their daughters will be carriers.
   (B) All their sons will be normal, and all their daughters will be color-blind.
   (C) All their sons will be color-blind, and all their daughters will be normal.
   (D) All their sons will be color-blind, and all their daughters will be carriers.
   (E) All their daughters will be carriers, but the genotype of the sons cannot be determined.

10. The terms blowout, pioneer organisms, and climax community are most related to a discussion of

    (A) biomes
    (B) ecological succession
    (C) food pyramids
    (D) populations
    (E) food chains

11. In humans, the traits for height, skin color, and hair color vary widely. The inheritance patterns of these traits is referred to as

    (A) polygenic inheritance
    (B) incomplete dominance
    (C) codominance
    (D) multiple alleles
    (E) simple inheritance

12. All of the following are metabolic wastes EXCEPT

    (A) water
    (B) carbon dioxide
    (C) uric acid
    (D) urea
    (E) oxygen

13. From the left ventricle, blood flows into the

    (A) pulmonary artery
    (B) aorta
    (C) left atrium
    (D) vena cava
    (E) septum

Directions: Each set of lettered choices below refers to the numbered questions or statements immediately following it. Select the one lettered choice that best answers each question and fill in the corresponding space on the answer sheet. A choice may be used once, more than once, or not at all in each set.

Questions 14–18

    (A) Light-dependent reactions
    (B) Light-independent reactions
    (C) Krebs cycle
    (D) Glycolysis

14. Occurs in cytoplasm

15. Water is broken down

16. Pyruvate is the product

17. Sugar is produced

18. Oxygen is released

Questions 19–22

    (A) Marine biome
    (B) Desert
    (C) Taiga
    (D) Temperate grasslands
    (E) Tundra

19. Has the most stable temperatures

20. Northern Canada—characterized by coniferous trees, such as spruce and fir

21. Plains and prairies of the midwestern United States

22. Permafrost

Questions 23–25

In a certain population of rabbits in the midwestern United States, there are two alleles for coat color, brown and white. Brown is dominant, and white is recessive. The frequency of white-colored rabbits is 9%. Wolves are also present in the area, and rabbits constitute a major portion of their diet. Foxes recognize prey when they do not blend in to the environment.

23. What is the frequency of the allele for brown coat color in the rabbits?

    (A) 0.09
    (B) 0.03
    (C) 0.3
    (D) 0.7
    (E) 0.9

24. In this population of rabbits, what percentage is hybrid brown?

    (A) 7%
    (B) 42%
    (C) 49%
    (D) 81%
    (E) 91%

25. If the climate were to change so that snow covered the ground much of the time, what change in the population of rabbits would you expect?

    (A) The frequency of the white allele would increase.
    (B) The frequency of the white allele would decrease.
    (C) The frequency of the brown allele would increase.
    (D) The population of wolves would decrease and then increase.
    (E) The population of wolves would increase and then decrease.

Questions 26–28

An experiment was carried out to study the rate of transpiration in plants. Four two-week-old bean seedlings were used. Each one was inserted into one end of a length of plastic tubing, and a calibrated pipette was inserted into the other end as shown below. The tubing was completely filled with water. As the plant loses water from its leaves by transpiration, it takes in more water from the pipette. The quantity of water absorbed by the plant can be measured by taking readings from the pipette. See drawing.

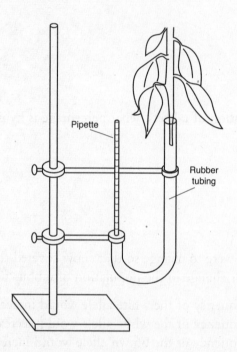

Each of the four setups is exposed to different conditions. The experiment is allowed to run for 2 hours with readings taken every 15 minutes. The control setup is at room temperature with normal room illumination. Predict which line on the graph matches each condition below:

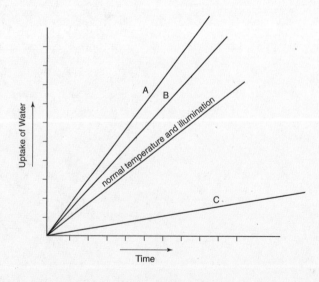

26. A fan is placed so that it circulates the air around the plant

27. A plastic bag is placed over the leaves

28. A second light source is added

Questions 29–32

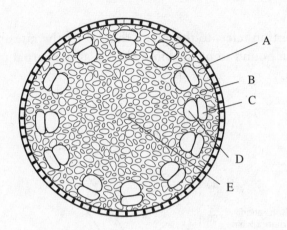

29. This is a

    (A) dicot stem
    (B) dicot root
    (C) monocot stem
    (D) monocot root
    (E) It cannot be determined

30. Structure C

    (A) carries water downward
    (B) assists in the absorption of nutrients into the plant
    (C) carries water upward
    (D) carries sugar upward
    (E) carries sugar downward

31. Structure E is the

    (A) meristem
    (B) cambium
    (C) vascular cylinder
    (D) pith
    (E) cortex

32. The main function of E is

   (A) transport
   (B) photosynthesis
   (C) storage
   (D) protection
   (E) absorption of nutrients

Questions 33–34

The following question refers to this graph that shows the rate at which relative amounts of oxygen bound to hemoglobin change as the partial pressure of oxygen changes.

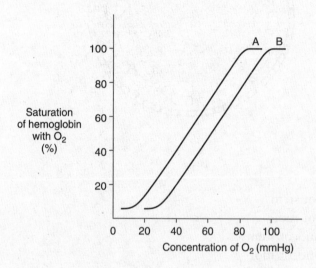

33. Based on this graph, which of the following statements is true?

   (A) At oxygen concentration 40 mm Hg, hemoglobin A is 40% saturated with oxygen, while hemoglobin B is at 20% saturation.
   (B) At oxygen concentration 60 mm Hg, hemoglobin B is 100% saturated with oxygen.
   (C) Hemoglobin A becomes 100% saturated with oxygen at 60 mm Hg.
   (D) Hemoglobins A and B become saturated with oxygen at the same concentration of O_2.
   (E) Neither hemoglobin A nor hemoglobin B ever becomes 100% saturated with oxygen.

34. According to the graph, what generality can be drawn about hemoglobin A or B?

(A) Hemoglobin B has a greater affinity for oxygen.
(B) Hemoglobin B would more easily drop off oxygen at body cells than hemoglobin A.
(C) Hemoglobin B would more easily pick up oxygen in the lungs than hemoglobin A.
(D) Hemoglobin A is structurally identical to hemoglobin B; however, they function differently.
(E) Hemoglobin A is found in adults, while hemoglobin B is found in fetuses and newborns only.

35. Sexually reproducing organisms have the diploid number ($2n$) of chromosomes. This is an advantage in terms of their evolutionary success. Which of the following best explains the reason for this?

I. The diploid condition hides tremendous genetic variation.
II. Diploid cells are superior to monoploid (n) cells.
III. The diploid condition allows for mutation, whereas the monoploid condition does not.

(A) I only
(B) II only
(C) III only
(D) I and III only
(E) I, II, and III only

Questions 36–40

Five cells are undergoing cell division.

   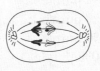

A   B   C   D   E

36. Crossing-over occurs

37. Sister chromatids are separating

38. Metaphase I

39. Homologous pairs are separating

40. Prophase I

<u>Questions 41–43</u>

Refer to this diagram of the female reproductive system.

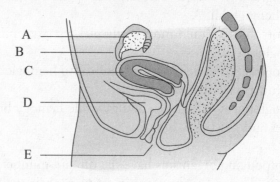

41. Fertilization normally occurs

42. Implantation of an embryo normally occurs

43. Urine is stored

<u>Questions 44–47</u>

In pea plants, the purple flower is an autosomal trait that is dominant over the white flower. Breeding experiments were carried out in order to study this trait. The choices below represent the percentages of purple-flowered plants in the first-generation offspring produced by different crosses that were made during the breeding experiments.

(A) 0% purple
(B) 25% purple
(C) 50% purple
(D) 75% purple
(E) 100% purple

44. Result from crosses between two hybrid purple pea plants

45. Result from crosses between a homozygous purple flower and homozygous white flower

46. Result from crosses between a purple plant (which had a white parent) and a white plant

47. Result from crosses between a hybrid purple plant and a white plant

Questions 48–50

This graph represents information about survival or mortality rates for three different populations.

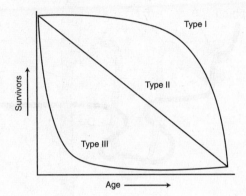

(A) Type I only
(B) Type II only
(C) Type III only
(D) Types I and III only
(E) Types I, II, and III

48. Organisms that carry out external fertilization and development

49. Include mammals

50. Experience enormous predation of young

Questions 51–53

(A) Cnidarians
(B) Chordates
(C) Annelids
(D) Roundworms
(E) Flatworms

51. Radial symmetry

52. Two cell layers thick

53. Three cell layers thick but have no coelom

Questions 54–56

Refer to this diagram of the nephron.

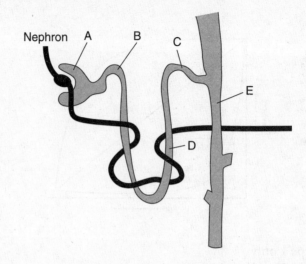

54. Controlled by a hormone from the posterior pituitary

55. Site of filtration

56. Site of reabsorption of nutrients back into the blood by active and passive transport

Questions 57–60

    (A) Adrenal cortex
    (B) Adrenal medulla
    (C) Pancreas
    (D) Thymus
    (E) Anterior pituitary

57. Important to the development of the immune system

58. Secretes hormones that stimulate the ovaries and testes

59. Both increases and decreases blood sugar levels

60. Secretes the fight-or-flight hormone

**If you are taking the Biology-E test, continue with questions 61–80.
If you are taking the Biology-M test, go to question 81 now.**

Biology-E Section

If you are taking the Biology-E test, continue with questions 61–80. Be sure to begin this section of the test by filling in circle 61 on your answer sheet.

61. All of the following are density-independent factors EXCEPT

    (A) floods
    (B) famine
    (C) earthquakes
    (D) tsunamis
    (E) naturally occurring fires

62. What is the role of diatoms in any food chain?

    (A) Primary consumers
    (B) Secondary consumers
    (C) Tertiary consumers
    (D) Producers
    (E) Decomposers

63. Two distinctly different beak sizes occur in a single population of finch called the black-bellied seedcracker. These birds live in an isolated region in West Africa. The oldest inhabitants of the region remember that all these finches used to have the same length beak. This change in the population is shown by this graph. The best explanation for the change in beak length is

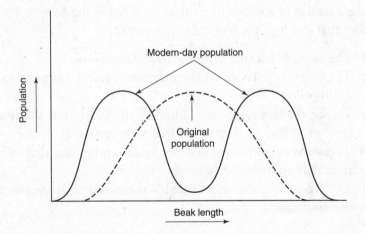

    (A) mutation
    (B) stabilizing selection
    (C) convergent evolution
    (D) genetic drift
    (E) diversifying selection

64. All of the following statements about Earth's ozone layer are correct EXCEPT

    (A) it is composed of O_3
    (B) chlorofluorocarbons destroy the ozone layer
    (C) the ozone layer protects us from skin cancer
    (D) it shields Earth from too much ultraviolet radiation
    (E) its thickness is expected to remain fairly constant for the next 100 years

65. The world human population today can best be described as

    (A) fluctuating around the carrying capacity
    (B) at equilibrium
    (C) growing arithmetically
    (D) growing exponentially
    (E) growing very slowly

66. Which of the following regions would contain the greatest diversity of species?

    (A) The Great Lakes
    (B) Costa Rica
    (C) Texas
    (D) Central Canada
    (E) The Mississippi Delta

67. After taking a one-week course of an antibiotic for an ear infection, a woman was told that she still had the infection and that she would have to take a course of a different antibiotic. What is the best explanation for the fact that the first antibiotic did not work?

    (A) The woman became immune to the antibiotic.
    (B) The bacteria that caused the infection became immune to the antibiotic.
    (C) Taking the first antibiotic killed off all the bacteria that were susceptible to it, leaving only those that were resistant.
    (D) The woman contracted another infection because there are no bacteria that are resistant to antibiotics.
    (E) The woman's hormones probably interfered with the functioning of the antibiotic.

68. Which of the following constitutes a likely food chain?

    (A) Squid—small fish—algae—orca whales—copepods
    (B) Copepods—squid—small fish—algae—orca whales
    (C) Algae—copepods—small fish—squid—orca whales
    (D) Copepods—algae—squid—small fish—orca whales
    (E) Algae—small fish—squid—copepods—orca whales

69. Characteristics of the arctic tundra biome include which of the following?

    I. Lichens growing on bare rock
    II. High levels of precipitation
    III. Permafrost

    (A) I only
    (B) III only
    (C) I and II only
    (D) I and III only
    (E) I, II, and III

70. Deep-ocean communities (abyssal zone) are dimly lit and have very few photosynthetic organisms. Which of the following is true of these areas?

    (A) Primary productivity is high
    (B) Levels of dissolved oxygen are low
    (C) Contain most of the organisms in the oceans
    (D) High concentration of algae
    (E) High concentration of green plants

Questions 71–72

The table below indicates the output of nitrate in stream water in an undisturbed area and in an area that has been clear-cut (deforested).

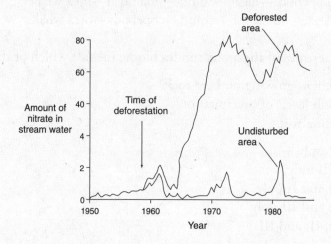

71. Which statement is indicated by the graph?

    (A) Nitrate runoff from the undisturbed area reached a maximum around 1973
    (B) Deforestation did not affect nitrate runoff for at least 10 years
    (C) Runoff rates were about the same until about 1958
    (D) Clear-cut occurred in about 1968
    (E) Maximum nitrate runoff in the clear-cut area occurred in 1978

72. Which statement about deforestation can be inferred from the graph?

    (A) Deforestation ultimately destroyed several small rivers in Oregon.
    (B) When an area is clear-cut, the trees can be replaced and the ecology of the area can be preserved.
    (C) The presence of trees in a forest causes an increase in nitrogen in the soil.
    (D) Plants and their root fungi are very efficient at absorbing fixed nitrogen from the soil.
    (E) Nitrogen runoff will destroy local rivers.

Questions 73–75

Refer to the following processes.

(A) Ecological succession
(B) Commensalism
(C) Eutrophication
(D) Evolution
(E) Adaptive radiation

73. Due to runoff of farm nutrients, the plant population blooms and the pond eventually fills in and becomes a terrestrial habitat

74. Grasses and shrubs give way to fast-growing trees

75. Due to competition for leaves on tall acacia trees, only the long-necked giraffes survived

Questions 76–77

Refer to the climate graph below.

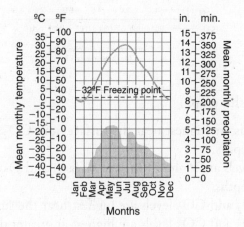

76. Which environmental condition(s) are shown on this graph?

(A) Alternating wet and dry seasons; temperatures that spike in summer and dip very low in fiercely cold winters
(B) Cold temperatures in the summer with warm temperatures in the winter; precipitation varies from month to month
(C) High temperatures most of the year with dry summer months and wet winter months
(D) High temperatures in the summer; fiercely cold winter with little or no rainfall most months
(E) High even temperatures all year; receives about 200 mm of rain per month all year long

77. Which biome is described by this graph?

    (A) Tropical rain forest
    (B) Arctic tundra
    (C) Taiga
    (D) Desert
    (E) Grasslands

Questions 78–80

Refer to this graph that shows the concentration of dissolved gas (parts per million by weight) at different depths in the ocean.

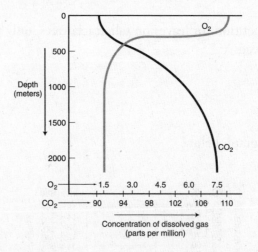

78. According to this graph,

    (A) O_2 levels are highest near the surface, and CO_2 levels are highest at great depths.
    (B) CO_2 levels are highest near the surface, and O_2 levels are highest at great depths.
    (C) Both O_2 and CO_2 levels are highest near the surface.
    (D) Both O_2 and CO_2 levels are highest at greater depths.
    (E) The concentrations of the gases at different depths change with the seasons.

79. According to the graph, what is the concentration of oxygen at the surface of the ocean?

    (A) 1.5 ppm
    (B) 4.5 ppm
    (C) 7.5 ppm
    (D) 98 ppm
    (E) 110 ppm

80. What is the correct explanation for the concentration of gases at different depths?

(A) Food chains are more elaborate at great depths in the oceans.

(B) Near the surface, animals use up O_2; while at great depths, animals produce large amounts of CO_2.

(C) Near the surface, great activity by aerobic organisms produces large amounts of CO_2; while at great depths there is little animal life, so little CO_2 is released.

(D) Near the surface, plants produce O_2; while below the sunlit layer, respiration of aerobes and decomposers uses up O_2 and releases CO_2.

(E) Cannot be determined given the information.

STOP

If you finish before time is called, you may check your work on the entire biology test only. Do not turn to any other test in the book.

Biology-M Section

If you are taking the Biology-M test, continue with questions 81–100. Be sure to begin this section of the test by filling in circle 81 on your answer sheet.

81. Which tool would be best to study the internal structure of an animal cell?

    (A) Transmission electron microscope
    (B) Scanning electron microscope
    (C) Phase-contrast microscope
    (D) Simple light microscope
    (E) X-ray diffraction

82. All of the following about plasma membrane structure and function are correct EXCEPT

    (A) diffusion of gases across a membrane require that the membrane be moist
    (B) all plasma membranes have the identical composition and structure
    (C) facilitated diffusion is an example of passive transport
    (D) proteins serve as membrane channels
    (E) plasma membranes contain receptors that are specific for the molecules they uptake

83. Vitamins are essential for normal cell function. They are important because they

    (A) function as an energy source
    (B) are hormones
    (C) directly assist in the normal conduction of impulses
    (D) resist pH changes
    (E) enable enzymes to function normally

Questions 84–85

Refer to this sketch of a cell undergoing meiosis to produce a gamete.

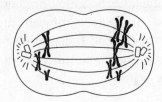

84. Which term best describes what is happening to this dividing cell?

   (A) Crossing-over
   (B) Metaphase spread
   (C) Gene mutation
   (D) Nondisjunction
   (E) Mitotic cell division

85. Which of the following conditions could result if a normal sperm fertilized one of the gametes produced by this cell?

   (A) Sickle cell anemia
   (B) Down syndrome
   (C) Cystic fibrosis
   (D) Huntington's disease
   (E) PKU

86. Here is a drawing of a field of view of a light microscope. Which way do you move the slide in order to move the cell into the center of the field?

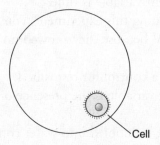

Cell

   (A) To the right and up
   (B) To the right and down
   (C) To the left and up
   (D) To the left and down

Questions 87–89

Four test tubes are required for this experiment. All the tubes contain pond water and the pH indicator bromothymol blue, which turns yellow when acidic and blue in the neutral or basic range. Carbon dioxide is bubbled into all four test tubes, causing the solutions to turn yellow. Sprigs of the plant elodea are placed into test tubes I and III. Test tubes I and II are kept in the light; test tubes III and IV are kept in the dark. After 24 hours, all tubes are examined and the observations are recorded in the table below.

Test Tube	Color at Time Zero	Color After 24 Hours
I	Yellow	Blue
II	Yellow	Yellow
III	Yellow	Yellow
IV	Yellow	Yellow

87. After 24 hours, tube I turned blue because the

    (A) plant excreted a by-product of cellular respiration, which has pH of 10
    (B) plant excreted carbon dioxide as a result of carrying out cellular respiration
    (C) plant excreted oxygen as a result of carrying out cellular respiration
    (D) plant excreted oxygen as a result of carrying out photosynthesis
    (E) plant absorbed carbon dioxide to use in photosynthesis

88. Which of the following is correct?

    (A) The independent variable is pH.
    (B) The independent variable is time.
    (C) If you were redoing this experiment, you would not have to use test tubes III and IV because they showed no change and are therefore not necessary.
    (D) Test tube III is a control for test tube I.
    (E) The dependent variable is the presence of oxygen.

89. When carbon dioxide is bubbled into water containing bromothymol blue,

    (A) the pH increases
    (B) the pH decreases
    (C) the water becomes alkaline
    (D) the water turns blue
    (E) there is no change in pH because carbon dioxide is a pH buffer

Questions 90–93

Students do an experiment to explore the effects of pH on enzyme activity. They use the enzyme catalase, which is found in all aerobic cells. The function of catalase is to breakdown the harmful waste product of cellular respiration, hydrogen peroxide (H_2O_2) into oxygen gas and water. The reaction looks like this: $2H_2O_2 \rightarrow 2H_2O + O_2$

The students set up 10 tests tubes, each containing 5 mL of peroxide. They adjust the pH of each test tube as shown below. To test tubes #1–5 they add 3 mL of catalase. To test tubes #6–10, they add 3 mL of distilled water. Immediately, some of the tubes begin to bubble. The students use a glowing splint to test for the presence of oxygen and to prove that the bubbles contain oxygen gas. Their data is in the chart below.

Test Tube	pH	Contents	Amount of Bubbling
1	1	3 mL Catalase	None
2	3	3 mL Catalase	+
3	5	3 mL Catalase	+++
4	7	3 mL Catalase	+++++
5	9	3 mL Catalase	+
6	1	3 mL water	None
7	3	3 mL water	None
8	5	3 mL water	None
9	7	3 mL water	None
10	9	3 mL water	None

90. According to the data shown, catalase is most effective under which of the following conditions?

 (A) In any acid medium
 (B) In the tubes that contain water
 (C) At pH 1
 (D) At pH 7
 (E) At pH 9

91. This experiment lends support to the idea that enzymes

 (A) Function within a narrow pH range
 (B) Function best at high pH
 (C) Function best at pH 6–7, which is the pH of the average cell
 (D) Function differently in a test tube from how they function in the body
 (E) Denature at high temperatures

92. What is the reason for adding water to test tubes #6–10?

    (A) The water is necessary in order to adjust the pH
    (B) The water replaces enzyme in those tubes
    (C) The water is necessary to dissolve the catalase
    (D) Water enhances the reactivity of catalase
    (E) Water is necessary to limit wide temperature changes

93. Which organelle in an animal cell produces the most hydrogen peroxide (H_2O_2)?

    (A) Nucleus
    (B) Ribosome
    (C) Nucleolus
    (D) Golgi body
    (E) Mitochondria

Questions 94–96

A tissue or cells can be mashed up in a blender to form a liquid homogenate and then spun in an ultracentrifuge at high speeds to separate it into layers based on differences in density. The densest cell organelles settle to the bottom of the tube, forming a pellet. The supernatant, the liquid above the pellet, which contains less-dense organelles, can be poured off and respun. In this way, cell organelles can be isolated. A homogenate of human tissue was processed and spun in this way. See the diagram below.

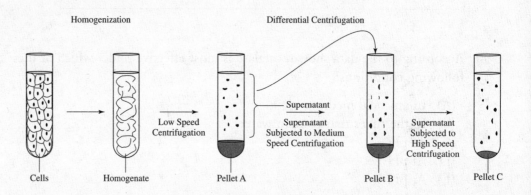

94. Pellet A was the first one to be separated by the centrifuge. When analyzed, it was found to consist of one cell organelle. Which one?

    (A) Nuclei
    (B) Mitochondria
    (C) Vacuoles
    (D) Golgi
    (E) Ribosomes

95. Pellet B was found to have the highest rate of oxygen uptake compared with the other pellets. Which organelles did it contain?

    (A) Ribosomes
    (B) Mitochondria
    (C) Vacuoles
    (D) Golgi
    (E) Nuclei

96. Pellet C was analyzed and found to consist of the highest concentrations of RNA compared with the other pellets. Which organelles did it contain?

    (A) Ribosomes
    (B) Mitochondria
    (C) Vacuoles
    (D) Golgi
    (E) Endoplasmic reticulum

Questions 97–100

Refer to this drawing of fertilization and the early embryonic development of a zygote.

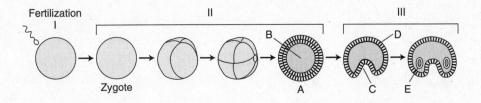

97. The process shown at II is called

    (A) morphogenesis
    (B) gastrulation
    (C) cleavage
    (D) meiosis
    (E) differentiation

98. The name of the hollow stage at A is

    (A) early embryo
    (B) morula
    (C) gastrula
    (D) blastula
    (E) deuterostome

99. Which stage is characterized by monoploid cells?

    (A) I
    (B) II
    (C) III

100. The structures at III result from

    (A) transcription
    (B) translation
    (C) budding
    (D) many mitotic divisions
    (E) many meiotic divisions

STOP

Determine Your Raw Score

Step 1: Compare your answer sheet to the correct answers on Table 21.1.

- Put a check in the column marked "Right" if your answer is correct.
- Put a check in the column marked "Wrong" if your answer is incorrect.
- Leave both columns blank if you omitted the question.

Step 2: Count the number of right answers, and enter the
number here: _____

Step 3: Count the number of wrong answers, divide by 4, and enter
the number here: _____

Step 4: Subtract the number you obtained in Step 3 from the number
in Step 2. Round the result to the nearest whole number
(0.5 is rounded up), and enter here: _____

Step 5: The number you obtained in Step 4 is your raw score.
Convert it to your College Board Score using Tables 21.2
and 21.3, which are similar to those published by The
College Board.

TABLE 21.1

The Correct Answers

Question No.	Correct Answer	Right	Wrong	Question No.	Correct Answer	Right	Wrong
1	E			26	A		
2	C			27	C		
3	D			28	B		
4	E			29	A		
5	B			30	E		
6	E			31	D		
7	D			32	C		
8	C			33	A		
9	A			34	B		
10	B			35	A		
11	A			36	B		
12	E			37	C		
13	B			38	E		
14	D			39	D		
15	A			40	B		
16	D			41	B		
17	B			42	C		
18	A			43	D		
19	A			44	D		
20	C			45	E		
21	D			46	C		
22	E			47	C		
23	D			48	C		
24	B			49	A		
25	A			50	C		

TABLE 21.1 (continued)

The Correct Answers

Question No.	Correct Answer	Right	Wrong	Question No.	Correct Answer	Right	Wrong
51	A			76	A		
52	A			77	E		
53	E			78	A		
54	E			79	C		
55	A			80	D		
56	D			81	A		
57	D			82	B		
58	E			83	E		
59	C			84	D		
60	B			85	B		
61	B			86	B		
62	D			87	E		
63	E			88	D		
64	E			89	B		
65	D			90	D		
66	B			91	A		
67	C			92	B		
68	C			93	E		
69	D			94	A		
70	B			95	B		
71	A			96	A		
72	D			97	C		
73	C			98	D		
74	A			99	A		
75	D			100	D		

TABLE 21.2

Score Conversion Table Biology-E

Raw Score	Scaled Score	Raw Score	Scaled Score	Raw Score	Scaled Score
80	800	50	630	20	440
79	800	49	620	19	430
78	800	48	610	18	430
77	800	47	600	17	420
76	800	46	590	16	410
75	800	45	590	15	410
74	800	44	580	14	400
73	800	43	580	13	390
72	790	42	570	12	390
71	780	41	570	11	380
70	780	40	560	10	380
69	770	39	560	9	370
68	760	38	550	8	360
67	750	37	550	7	360
66	750	36	540	6	360
65	740	35	540	5	350
64	730	34	530	4	340
63	720	33	530	3	330
62	710	32	520	2	330
61	700	31	510	1	320
60	690	30	500	0	310
59	690	29	490	−1	300
58	680	28	490	−2	290
57	670	27	480	−3	290
56	670	26	470	−4	280
55	660	25	470	−5	270
54	660	24	460	−6	270
53	650	23	460	−7	260
52	650	22	450	−8	250
51	640	21	440	−9	250

TABLE 21.3

Score Conversion Table Biology-M

Raw Score	Scaled Score	Raw Score	Scaled Score	Raw Score	Scaled Score
80	800	50	630	20	440
79	800	49	620	19	430
78	800	48	610	18	430
77	800	47	600	17	420
76	800	46	590	16	410
75	790	45	590	15	410
74	780	44	580	14	400
73	780	43	580	13	390
72	770	42	570	12	390
71	770	41	570	11	380
70	760	40	560	10	380
69	760	39	560	9	370
68	750	38	550	8	360
67	740	37	550	7	360
66	740	36	540	6	360
65	730	35	540	5	350
64	720	34	530	4	340
63	710	33	530	3	330
62	700	32	520	2	330
61	700	31	510	1	320
60	690	30	500	0	310
59	690	29	490	−1	300
58	680	28	490	−2	290
57	670	27	480	−3	290
56	670	26	470	−4	280
55	660	25	470	−5	270
54	660	24	460	−6	270
53	650	23	460	−7	260
52	650	22	450	−8	250
51	640	21	440	−9	250

Sample Test 2

EXPLANATION OF ANSWERS

1. **(E)** This cross is called a testcross or backcross. It is used to determine the genotype of an organism that shows the dominant phenotype but whose genotype is unknown. By crossing the organism with the recessive, you maximize the chance of getting any offspring with the recessive genotype (white). If the organism of unknown genotype is hybrid, the chance of getting a recessive offspring is 50%. Here is the cross.

	B	b
b	Bb	bb
b	Bb	bb

   If you cross the animal in question with another black animal that is also hybrid (choice (D)), the chance of getting a white offspring is only 25%. Here is that cross.

	B	b
B	BB	Bb
b	Bb	bb

   Choice (E) offers the best opportunity to get a white animal if the black parent is hybrid. A karyotype would reveal if there is any nondisjunction in the chromosomes, just as choice (B) describes. That does not help at all because genes (for fur color) cannot be seen by looking at chromosomes. Using any type of microscope would not help either, because the inheritance of fur color is simple dominance and the fur of a hybrid black individual looks exactly the same as the fur of a pure black individual.

2. **(C)** Fixed action patterns are innate, stereotypical animal behaviors that, once begun, must be continued to completion. They are initiated by sign stimuli. If the stimuli are exchanged between members of the same species, they are known as releasers.

3. **(D)** Lichens look like plants but are actually mutualistic symbionts consisting of fungi that hold the symbiont to a surface and algae, which are photosynthetic.

4. **(E)** Abiotic means nonliving. Producers are the only biotic factors in the list. An ecosystem depends on the interplay between biotic and abiotic factors.

5. **(B)** Nitrogen-fixing bacteria live in nodes on the roots of legumes. They convert free nitrogen into the ammonium ion, a compound that plants can use to make amino acids. As a result, legumes are high in protein.

6. **(E)** In terms of evolution, single individuals never change. Those that are not adapted die or move away. Those that are best adapted survive and pass their (best-adapted) genes on to another generation. Only populations, groups of individuals, evolve. For example, one individual peppered moth never changed color. The entire population changed or shifted, because of pressure from the environment, from mostly light and few dark to mostly dark and few light.

7. **(D)** People with Down syndrome have 47 chromosomes because they have an extra 21st chromosome. This error is caused by nondisjunction. It is a chromosome mutation. The other choices are all mutations in a gene.

8. **(C)** Both the breathing and heart rate are controlled by the part of the brain known as the medulla or medulla oblongata. The word pleura relates to the lungs.

9. **(A)** All daughters of a man who has an X-linked condition will be carriers of the condition. Here is the cross:

	X–	Y
X	XX–	XY
X	XX–	XY

10. **(B)** All these terms are related to ecological succession, the natural process where there is a transition from one type of community to another after there is a disturbance of the ecosystem.

11. **(A)** When a trait varies along a continuum, such as skin color, it must be controlled by more than one gene. For example, there is only one gene that controls height in pea plants but it has two alleles: *T* (tall) or *t* (dwarf). In contrast, the trait for height in humans is controlled by several genes, each with more than one allele, and is referred to as polygenic inheritance. This is the source of the tremendous variability in height in humans.

    An example of multiple alleles is blood type. Although there is one gene for blood type, there are 3 alleles: *A*, *B*, and *O*. A person can have any one of these genotypes: *AA*, *AO*, *BB*, *BO*, or *AB*. Incomplete dominance is blending inheritance. An example of codominance is the case where both traits that are present express themselves. A person who has the blood alleles *A* and *B* has blood type AB.

12. **(E)** Oxygen is taken in, not given off, by aerobic animals to carry out respiration. The wastes from respiration are carbon dioxide and water vapor. The waste from protein metabolism is urea in humans, uric acid in birds, and ammonia in organisms that live in water.

13. **(B)** The left ventricle pumps blood to the entire body via the aorta. The atria are receiving chambers.

14. **(D)** Glycolysis occurs in the cytoplasm. The light-dependent reactions occur in the grana of the chloroplast. The light-independent reactions occur in the stroma of the chloroplast. The Krebs cycle occurs in the inner matrix of mitochondria.

15. **(A)** Water is broken down by the process called photolysis during the light-dependent reactions. Water breaks down when electrons are pulled away from it to replace those lost from chlorophyll *a*.

16. **(D)** Pyruvate is the product of glycolysis and it combines with coenzyme A to yield acetyl coA, which is the raw material for the Krebs cycle.

17. **(B)** Sugar is produced during the light-independent reactions (dark reactions). It occurs in the stroma of chloroplasts during carbon fixation by a process called the Calvin cycle.

18. **(A)** Oxygen from water molecules is released from the light-dependent reactions in the grana of chloroplasts. Water is broken down (photolysis) as electrons are pulled away to replace those that were lost from chlorophyll *a*. Once those electrons are removed, oxygen is released to the air.

19. **(A)** The marine biome is the largest and most stable biome. Temperatures vary little because water has a high heat capacity and there is such an enormous volume of water.

20. **(C)** The taiga is dominated by conifers. The landscape is dotted with lakes, ponds, and bogs, and it is the largest terrestrial biome. Winters are very cold, and the principal mammals are moose, elk, deer, and black bear.

21. **(D)** Temperate grasslands are characterized by low total annual rainfall or uneven seasonal rainfall. Gazing animals include bison and longhorn antelope in the United States. In Africa, the mammal inhabitants are wildebeest and gazelle.

22. **(E)** The permafrost is permanently frozen subsoil found in the farthest northern lands, including Alaska. Though the number of individual organisms, including reindeer, caribou, and polar bears is high, the number of species is low.

23. **(D)** In this example, 9% of the rabbits exhibit the recessive trait (white), which is known as q^2 according to Hardy-Weinberg. If q^2 = 9% or 0.09, then q = the square root of 0.09, which is 0.3. So if the allele for q is 0.3, then the allele for brown is 0.7 because $p + q = 1$ and $0.3 + 0.7 = 1$.

24. **(B)** In this example, 9% of the rabbits exhibit the recessive trait (white), which is known as q^2 according to Hardy-Weinberg. If q^2 = 9% or 0.09, then q = the square root of 0.09 or 0.3. If q equals 0.3, then $p = 0.7$ because $p + q = 1$. You are asked for the hybrid condition, which is $2pq$. So $2 \times 0.3 \times 0.7 = 0.42$ or 42%.

25. **(A)** If the climate changes and the ground is covered by snow more of the time, then rabbits that are better camouflaged will survive while animals that are not camouflaged will be killed and eaten by wolves. So animals with white fur will survive better than animals with brown fur, and the frequency of the allele for white fur will increase in the population.

26. **(A)** Circulating air around a leaf will increase the rate of transpiration to the greatest extent because it removes water vapor from the area around the stomates. The rate of transpiration increases as humidity in the air decreases.

27. **(C)** Placing a bag over the leaves will decrease the rate of transpiration drastically because it will increase the humidity around the leaf. The rate of transpiration decreases as the humidity in the air increases.

28. **(B)** Increasing the light will increase the rate of photosynthesis, which will increase the rate of transpiration. However, since the rate of transpiration is a function of humidity as well as the rate of photosynthesis, the humidity is a limiting factor. The rate of transpiration cannot increase as much as it did in question 26.

29. **(A)** In dicot stems, the vascular bindles are located in a ring around the edge. In monocot stems, they are scattered across the stem.

30. **(E)** Structure C is the phloem. It carries sugar around the plant, mostly downward from the leaves to the rest of the plant. The process of moving sugar around the plant is known as translocation.

31. **(D)** The pith stores starch. Pith cells are large, flexible, parenchymal cells.

32. **(C)** Pith stores starch. Pith cells are large, flexible, parenchymal cells.

33. **(A)** The *y*-axis of the graph shows the saturation of hemoglobin with oxygen. The *x*-axis shows the concentration of oxygen in mm Hg. If you look on the *x*-axis for 20 mm Hg and move up to the line for hemoglobin A and over from there to the *y*-axis, you see 40%. Also do the same thing for hemoglobin B. Hemoglobin B is saturated only to 20%.

34. **(B)** The further left the line is on the graph, the more easily hemoglobin binds to oxygen and the more difficult for that hemoglobin to drop oxygen at the cells. Since hemoglobin B binds less easily to oxygen (because it is to the right of A), hemoglobin B will more easily drop oxygen off at the cells. Structure relates to function and if the function is different, the structure cannot be the same.

35. **(A)** The diploid (2*n*) number of chromosomes hides tremendous variation because the recessive trait does not express itself. However, it can express itself in the next generation. This is an advantage that sexually reproducing organisms (2*n*) have over asexually reproducing (*n*) organisms. Both conditions can experience mutations. However, mutations will express themselves immediately in the monoploid condition (*n*). Diploid cells are not, in themselves, superior.

36. **(B)** The stage is prophase I when crossover occurs. Homologues chromosomes pair up during a process called synapsis and exchange homologous parts.

37. **(C)** This stage is anaphase II. Sister chromatids separate during meiosis II, while homologous chromosomes separate during meiosis I.

38. **(E)** This stage is metaphase I, when homologous pairs line up double file.

39. **(D)** This stage is anaphase I; homologous pairs are separating.

40. **(B)** This stage is prophase I. The nuclear membrane has disintegrated, and chromosomes condense and become visible under the light microscope.

41. **(B)** Fertilization normally occurs in the oviduct or Fallopian tube.

42. **(C)** Implantation of the embryo normally occurs in the uterus. Although fertilization occurs in the oviduct, the fertilized egg or zygote takes several days to travel to the uterus where it will implant.

43. **(D)** This is the urinary bladder where urine is temporarily stored.

44. **(D)** Purple (*P*) is dominant over white (*p*). Here is the cross.

	P	*p*
P	*PP*	*Pp*
P	*Pp*	*pp*

75% of the offspring will be purple; 25% will be white.

45. **(E)** The cross is between pure purple (*PP*) and white (*pp*). Here is the cross.

	P	*P*
p	*Pp*	*Pp*
p	*Pp*	*Pp*

All the offspring are hybrid purple.

46. **(C)** A purple plant (that had a white parent) *Pp* is crossed with a white plant (*pp*). Here is the cross.

	P	*p*
p	*Pp*	*pp*
p	*Pp*	*pp*

50% of offspring will be hybrid purple, and 50% will be white.

47. **(C)** This is the same cross as the one shown in question 46.

48. **(C)** The animals described by type III are characterized by enormous death of young, but the few young that survive do live a long life. Most young die immediately, which is probably due to predation. These animals most likely exhibit external fertilization and development, which would account for the enormous rate of death among the young.

49. **(A)** Type I describes animals with a high survival rate of the young because of intense parenting. However, adults seem to die in old age.

50. **(C)** The animals described by type III are characterized by enormous death of the young. This is most likely due to predation because eggs are fertilized and develop externally.

51. **(A)** Cnidarians are primitive animals that are characterized by radial symmetry.

52. **(A)** Cnidarians are primitive animals that are characterized by having two cell layers.

53. **(E)** Flatworms belong to the phylum Platyhelminthes. An example is the planaria. They are three cell layers thick, but they are acoelomates. They also exhibit a head region, cephalization.

54. **(E)** Structure E is the collecting duct, which is under hormonal control from the posterior pituitary—ADH, antidiuretic hormone. ADH controls the permeability of the membrane that controls the absorption of water back into the body. If ADH is blocked, more urine is excreted. If more ADH is released, less urine is released.

55. **(A)** The site of filtration is Bowman's capsule. Nutrients and wastes diffuse from the glomerulus into Bowman's capsule.

56. **(D)** Reabsorption is the process by which most of the water and solutes (glucose, amino acids, and vitamins) that initially entered the tubule during filtration are transported back into the capillaries and, thus, back to the body. The longer the loop of Henle, the greater the reabsorption of water. Mammals that live in the Sonoran Desert in Arizona have longer loops of Henle than their counterparts that evolved in wetter climates where dehydration is not such a severe problem.

57. **(D)** The thymus is critical to the development of the immune system. It is active and large in the fetus and early childhood and becomes small and inactive as a person grows older. In adults, its function is not understood.

58. **(E)** The anterior pituitary is located in the brain and receives stimulation from the hypothalamus. It secretes hormones that stimulate other glands. These are FSH and LH—which stimulate the ovaries and testes, ACTH—

which stimulates the adrenal cortex, and HGH—which stimulates bone growth.

59. **(C)** The pancreas contains the islets of Langerhans, which secrete glucagon to raise blood sugar and insulin to lower it.

60. **(B)** The adrenal medulla secretes adrenalin, the "fight or flight" hormone. At times of extreme stress, it causes a sudden increase in heart rate, blood flow, and breathing rate. It also causes other parts of the body to be in a heightened state of readiness.

61. **(B)** Famine is a density-dependent factor resulting from overpopulation.

62. **(D)** Diatoms are among the most abundant organisms on Earth, making up the phytoplankton. They are photosynthetic organisms found near the surface of the oceans that carry out about half of all the photosynthesis on Earth. Diatoms belong to the phylum Protista in the subdivision Algae. They are delicate and as varied as snowflakes, and they have unique, glasslike walls consisting of silicon.

63. **(E)** Because of selective pressure from the environment, this population of birds divided into two distinct phenotypes. This is called diversifying or disruptive selection.

64. **(E)** Choices (A) through (D) are correct statements, but the ozone layer is being depleted. Beginning in the 1970s, scientists found that there was a hole in the ozone layer above Antarctica in winter. Over the last 30 years, the hole has been growing larger and lasting longer. The United States has joined with other nations to ban the use of CFCs (chlorofluorocarbons) that cause the destruction of the ozone layer.

65. **(D)** The human population is growing exponentially.

66. **(B)** Costa Rica contains a tropical rain forest that is a hotspot of species diversification. If not preserved, many organisms will become extinct.

67. **(C)** If you chose choice (B), you were careless. Remember, individual organisms do not change. Only populations change in response to pressure from the environment. The first antibiotic that the woman took killed off all the bacteria that were susceptible to it. However, some bacteria happened to be resistant to the antibiotic, and their population bloomed. In effect, exposure to an antibiotic selects for a resistant strain because it kills off the vulnerable strains.

68. **(C)** Copepods are small crustaceans that are among the most numerous of animals. They are important members of the freshwater and marine plankton communities, eating algae and themselves being eaten by small fish.

69. **(D)** The tundra is characterized by bitterly cold temperatures, permafrost, high winds (responsible for the absence of trees), and very little rainfall.

70. **(B)** The abyssal zone gets very little light and therefore has very few producers. Therefore, oxygen levels are low.

71. **(A)** According to the graph, the time of deforestation occurred in about 1958. Until about 1964, runoff rates were about the same. After that, runoff rates for the deforested area increased greatly, reaching a maximum around 1973.

72. **(D)** The graph does not name a locale or river. Although you may know that the other statements are actually true, they cannot be inferred from the graph.

The only one that can be inferred is choice (D). Apparently the presence of trees in a forest captures nitrogen in the soil and prevents it from running off into the rivers. The roots and fungi that live in the roots as mutualistic symbionts uptake and hold the nitrogen.

73. **(C)** The process of eutrophication will cause the disappearance of the lake by the accumulation of decomposed matter and humus.

74. **(A)** Ecological succession is a natural process that leads to a climax community like a deciduous forest or other biome. A major storm or flood can wash out or destroy a climax community in what is called a blowout.

75. **(D)** This is an evolutionary process. Adaptive radiation refers to the emergence of numerous species from one common ancestor introduced into an environment. An example is Darwin's finches on the Galapagos Islands. Commensalism is a form of symbiotic relationship where one organism benefits while the other is neither helped nor harmed.

76. **(A)** The line represents temperature, and the shaded-in area at the bottom represents precipitation.

77. **(E)** Grasslands are usually located in the interior, away from the moderating effects of the oceans, as in the central United States. Summers are hot and winters are very cold. The most rain is in the spring and early summer, but there can be long droughts. There is inadequate rain to support tall tree growth.

78. **(A)** Two lines are shown. The lighter line represents O_2, and the dark line represents CO_2. Choice (E) does not make sense because there is no information on the graph about seasonal changes.

79. **(C)** This is a read-off from the graph. At zero depth (on the surface), the concentration of oxygen is 7.5 ppm.

80. **(D)** Photosynthetic organisms at the surface (where there is sunlight) take in carbon dioxide and give off oxygen. At great depths, there is little sunlight and little photosynthesis but many aquatic organisms and decomposers that use oxygen and release carbon dioxide.

81. **(A)** Electron microscopes have the highest magnification and best resolution, so they would be best for studying tiny organelles. The transmission electron microscope enables us to see inside cells. The scanning electron microscope is best for studying surfaces of cells, not internal structures.

82. **(B)** Remember that form relates to function. Different cells with different functions would have membranes with different characteristics. The membrane of a bone cell must be very different from the membrane of a neuron.

83. **(E)** Vitamins act as coenzymes. Minerals are cofactors. Both assist enzymes in functioning normally.

84. **(D)** Nondisjunction is an abnormal occurrence. In this example, two gametes will form; one will have an extra chromosome while one will be lacking one chromosome.

85. **(B)** Down syndrome is a condition in which a person has an extra 21st chromosome. This chromosome mutation is caused by nondisjunction. The other 4 choices do not result from chromosomal abnormalities but from an error in a gene.

86. **(B)** Everything in the field of view is upside-down and backward.

87. **(E)** Tube I is yellow because carbon dioxide was bubbled into it. As the elodea carries out photosynthesis, it uses up the carbon dioxide, and the color of the bromothymol blue turns back to blue.

88. **(D)** Tube III is a control for tube I because the setup is exactly the same for both except that tube I is placed in the light and tube III is placed in the dark. The variable that changes in response to something that you do in the experiment is called the dependent variable. Time is the same for all tubes and is not relevant. The pH does not have anything to do with this experiment.

89. **(B)** Carbon dioxide dissolves in water to produce carbonic acid, so the pH decreases; it becomes more acidic.

90. **(D)** The amount of bubbling is a function of the extent of the reaction, the breakdown of H_2O_2 into H_2O and O_2. The most bubbling occurs at pH 7.

91. **(A)** Enzymes function best within a narrow pH range. Enzymes in the stomach work best at around pH 3; while intestinal enzymes functions best in a basic pH. While it is true that enzymes denature at high temperatures, that topic was not touched on in this experiment.

92. **(B)** The test tubes that contain water are the controls for each test tube that contains catalase.

93. **(E)** Catalase is the enzyme that breaks down hydrogen peroxide (H_2O_2), a toxic by-product of normal cellular respiration. Since mitochondria carry out cellular respiration, they produce the most H_2O_2.

94. **(A)** Processing tissue in this way is called cell fractionation. The densest particles settle out first and then the less dense ones and so on. Nuclei are the densest, and they settle out first.

95. **(B)** Mitochondria can carry out cell respiration and use up oxygen even though they are not inside a cell.

96. **(A)** Ribosomes consist of proteins and rRNA (ribosomal RNA).

97. **(C)** Cleavage is the rapid mitotic cell division that occurs immediately after fertilization.

98. **(D)** The hollow ball stage in cleavage is called the blastula. The solid ball stage is called the morula.

99. **(A)** Stage I includes the sperm and ova. These are monoploid (*n*) cells.

100. **(D)** Cleavage and gastrulation are the result of many mitotic divisions. The chromosome number in each cell remains unchanged.

What Topics Do You Need to Work On?

Table 21.4 shows an analysis by topic for each question on the test you just took.

TABLE 21.4

Topic Analysis				
Cellular and Molecular Biology	Heredity	Evolution and Diversity	Organismal Biology	Ecology
4, 14, 15, 16, 17, 18, 26, 27, 28, 36, 37, 38, 39, 40, 81–100	1, 7, 9, 11, 35, 44, 46, 47	6, 23, 24, 25, 51, 52, 53	2, 3, 8, 12, 13, 29, 30, 31, 32, 33, 34, 41, 42, 43, 48–50, 54, 55, 56, 57–60	5, 10, 19, 20, 21, 22, 61–80

APPENDIXES

Glossary

Abiotic factors Nonliving factors in an ecosystem. They include: temperature, water, sunlight, wind, rocks, and soil.

Abscisic acid (ABA) Plant hormone that inhibits growth.

Accessory pigments *See antennae.*

Acoelomate An animal that has no true coelom. Flatworms are an example.

Actin Protein that makes up the thin myofilaments in skeletal muscle.

Active transport Movement of particles against a gradient, from low concentration to high concentration. This always requires the expenditure of energy

Adenosine triphosphate *See ATP.*

Adventitious roots Roots that arise above ground; examples are aerial roots and prop roots.

Agonistic behavior Aggressive behavior.

Alcohol fermentation The process by which certain cells convert pyruvic acid or pyruvate from glycolysis into ethyl alcohol and carbon dioxide in the absence of oxygen.

Allantois Extra embryonic membrane in bird's egg. It exchanges respiratory gases to and from the embryo.

Alleles Alternate forms of a gene. For example, there are two alleles for height in pea plants, tall and dwarf.

Alveolus (alveoli, pl.) Microscopic air sacs in the lung where diffusion of the respiratory gases, oxygen and carbon dioxide occurs.

Amnion Membrane that encloses the embryo in protective amniotic fluid.

Amylase Enzyme that digests starch.

Analogous structures Structures, such as a bat's wing and a fly's wing, that have the same function but not the same underlying structure. The similarity is merely superficial and reflects adaptation to a similar environment. Analogous structures are not evidence of a common origin or common ancestry.

Aneuploidy Any abnormal condition of the chromosomes.

Angiosperms Anthophyta or flowering plants.

Anion A negative ion.

Antennae or accessory pigments Molecules that assist in photosynthesis by capturing and passing on photons of light to chlorophyll *a* and expanding the range of light that can be used to produce sugar. Examples are chlorophyll *b* and the carotenoids.

Anther Male part of flower where sperm (pollen) is produced by meiosis. Sits atop the filament.

Antheridia Structures located on the tips of a gametophyte plant and that produce sperm.

Antibodies Immunoglobins. Part of the third line of defense, the specific immune response. Each antibody molecule is a Y-shaped molecule consisting of four polypeptide chains.

Anticodon The three-nucleotide sequence associated with tRNA.

Antigens Anything that triggers an antibody response.

Apoptosis Programmed cell death.

Archaeopteryx An intermediate fossil that shows both reptile and bird characteristics.

Archegonia   Structures located on the tips of a gametophyte plant and that produce eggs.

Associative learning   One type of learning in which one stimulus becomes linked to another through experience.

ATP or **adenosine triphosphate**   Special high-energy molecule that stores energy for immediate use in the cell.

ATP synthetase   Structure in the membranes of mitochondria and chloroplasts where ATP is formed.

Autonomic nervous system   Part of the nervous system that controls automatic functions, such as heart and breathing rate.

Autosomes   Chromosomes other than the sex chromosomes. Humans normally have 44 in each body cell.

Autotrophs   Organisms that make their own food.

Auxins   Growth hormones in plants that are responsible for phototropisms and apical dominance, the preferential growth of a plant upward (toward the sun) rather than laterally.

Backcross   Testcross. A technique to determine whether an individual plant or animal showing the dominant trait is homozygous dominant (*BB*) or heterozygous (*Bb*).

Bacterial transformation   The ability of bacteria to alter their genetic makeup by uptaking foreign DNA from another bacterial cell and incorporating it into their own. Discovered by the scientist named Griffith.

Base-pair substitution   A mutation where one nucleotide is substituted for a correct one in the DNA strand.

Bicarbonate ion   The most important buffer in human blood. It is responsible for keeping the pH of the blood at 7.4.

Bile   Chemical produced in the liver and released from the gallbladder that emulsifies fats. It is not an enzyme.

Binomial nomenclature   System of taxonomy that we use today, developed by Carl von Linné. In this system, every organism has a two-part name, like *Homo sapiens.*

Biological magnification   Organisms at higher trophic levels have a greater concentration of accumulated toxins stored in their bodies than those at lower trophic levels.

Biosphere   The global ecosystem.

Biotechnology   The branch of science that uses recombinant DNA techniques for practical purposes, also called genetic engineering.

Biotic factor   Includes all the organisms with which an organism might react in an ecosystem.

Biotic potential   The maximum rate at which a population could increase under ideal conditions.

Bottleneck effect   Natural disasters such as fire, earthquake, and flood reduce the size of a population nonselectively, resulting in a loss of genetic variation. The resulting population is much smaller and not representative of the original one. Certain alleles may be under- or overrepresented compared with the original population.

Budding   Splitting off of new individuals from existing ones. How reproduction occurs in hydra.

Buffers   Chemicals that resist a change in pH.

C-4 photosynthesis   Modification for dry environments. C-4 plants exhibit modified anatomy and biochemical pathways, which enable them to minimize excessive water loss and maximize sugar production.

Calvin cycle   Cyclical process that produces sugar. It occurs during the light-independent reactions.

CAM   *See crassulacean acid metabolism.*

Carbon fixation   Incorporation of carbon dioxide into a sugar. It occurs during the cyclical process called the Calvin cycle.

Cardiac sphincter   Band of muscle at the top of the stomach that keeps acidified food in the stomach from backing up into the esophagus and burning it.

Carotenoids   Photosynthetic antennae pigments. They are orange and yellow.

Carpel   Female part of the flower, produce the female gametophytes, ova. Each carpel consists of an ovary, stigma, and style. Also called the pistil.

Carrying capacity (K)   A limit to the number of individuals that can occupy one area at a particular time.

Cation   A positive ion.

Centrioles   Responsible for division of the cytoplasm in animal cells; they are not present in plant cells. They consist of 9 triplets of microtubules arranged in a circle.

Centromere   Specialized region of a chromosome that holds two sister chromatids together.

Centrosome Consist of two centrioles at right angles to each other. Important during cell division in animal cells.

Chemiosmosis This is how ATP is produced during oxidative phosphorylation. Protons only flow through the special ATP synthetase channels and transfer energy to molecules of ATP.

Chitin A polysaccharide that makes up the exoskeleton of insects and the cell walls of fungi.

Chloroplasts Type of plastid that carries out photosynthesis.

Chorion Membrane that lies under the shell of an egg and allows for diffusion of respiratory gases between the outside environment and the inside of the shell.

Chromatin network DNA in the nucleus that is wrapped with special proteins called histones into a visible network.

Chromoplasts Type of plastid that stores pigments that are responsible for the bright colors in fruit and flowers.

Classical conditioning Type of associative learning. Pavlov trained dogs to associate the sound of a bell with food. The result of this conditioning was that dogs would salivate upon merely hearing the sound of the bell even though no food was present.

Cleavage Rapid mitotic cell division of the zygote that begins immediately after fertilization.

Cnidocytes Cells that house the stingers in cnidarians.

Codominance An inheritance pattern where both traits show at once. In humans, a person who has 2 different genes for blood type, A and B, has type AB blood.

Codon The three-nucleotide sequence associated with mRNA.

Coelomate An animal that has a true coelom. All chordates are coelomates.

Coenzymes Vitamins that assist in the normal functioning of enzymes.

Coevolution The mutual evolutionary set of adaptations of two interacting species.

Cofactors Minerals that assist in the normal functioning of enzymes.

Cohesion tension The attraction of like molecules to stick together. Water molecules tend to stick together because they exhibit strong cohesion tension.

Collenchyma cells Plant cells that have unevenly thickened cell walls but lack secondary cell walls. The strings of celery consist of collenchyma cells.

Colon Another name for large intestine.

Commensalism Symbiotic relationship in which one organism benefits and one is not affected by the other organism (+/o).

Community Consists of all the organisms living in one area.

Companion cells Make up phloem vessels, along with sieve tube elements.

Conjugation A primitive form of sexual reproduction where individuals exchange genetic material.

Continental drift The theory that states that the continents are floating and moving very slowly. Over millions of years, seven separate continents formed from one original continent, Pangea.

Contractile vacuole Structure found in freshwater protista, like paramecia and amoeba, that pumps out excess water that diffuses inward because the organisms live in an environment that is hypotonic.

Convergent evolution Type of evolution where unrelated species occupying the same environment and subjected to similar selective pressures show similar adaptations. The classic example is the whale (a mammal) and the fish.

Cortex Specialized region in a plant root or stem for storage and support.

Cotyledon Food for the growing embryo in a dicot seed. The cells that make up the cotyledon are triploid ($3n$).

Covalent bonds Bonds formed between atoms where electrons are shared.

Crassulacean acid metabolism or **CAM** A form of photosynthesis that is an adaptation for dry conditions. These plants keep their stomates closed during the day and open at night, the reverse of how most plants behave.

Cristae Series of inner membranes in mitochondria where cell respiration occurs.

Crop Structure in birds, insects, and earthworms, among others, for temporary storage of food.

Crossing-over A normal process in which homologous chromatids exchange genetic material. Crossover is important because it increases variation in the gametes.

Cutin   Waxy coating on the leaves that helps prevent excess water loss from the plant.

Cyclosis   Movement of cytoplasm around the cell.

Cystic fibrosis   The most common lethal genetic disease in the United States, 1 out of 25 Caucasians is a carrier. Characterized by build-up of extracellular fluid in the lungs and digestive tract.

Cytokinesis   Division of the cytoplasm. In animal cells, a cleavage furrow forms down the middle of the cell as the cytoplasm pinches inward and the two daughter cells separate from each other. In plant cells, a cell plate forms down the middle of the cell.

Cytokinins   Plant hormones that stimulate cell division and cytokinesis.

Cytoplasm   The entire region between the nucleus and plasma membrane.

Cytosol   Semiliquid portion of the cytoplasm.

Decomposer   Organisms that play a vital role in the ecosystem and that recycle dead organic matter. Examples are bacteria and fungi.

Dehydration synthesis   Also known as synthesis. Process by which molecules are bonded together to form a larger molecule with the removal of water.

Deletion   A mutation where a piece of a gene, or chromosome is lost.

Denature   Characteristic of proteins; a change in shape that stops the protein from functioning.

Deoxyribonucleic acid or **DNA**   The heritable material, passed from parent to offspring.

Diastole   Relaxation of the ventricles of the heart. Normal diastolic pressure is 120 mm Hg.

Dicotyledon   Plant whose seed easily breaks in two.

Diffusion   The flow of molecules from a higher concentration to a lower concentration. There are two types: simple and facilitated.

Digestion   Enzymatic breakdown, hydrolysis, of food so it is small enough to be assimilated into the body.

Dipeptide   A molecule consisting of two amino acids.

Directional selection   Changing environmental conditions give rise to this type of natural selection. One phenotype replaces another in the gene pool.

Disruptive selection   This type of natural selection increases the numbers of extreme types in a population at the expense of intermediate forms.

Divergent evolution   Occurs when a population becomes isolated (for any reason) from the rest of the species and becomes exposed to new selective pressures, causing it to evolve into a new species. Homologous structures are evidence of divergent evolution.

DNA   *See deoxyribonucleic acid.*

DNA polymerase   The enzyme that catalyzes the elongation of the new DNA strands during replication.

Domain   In the newest system of classification, all organisms are classified in one of three domains, which are further divided into kingdom, phylum, class, order, family, genus, and species.

Duodenum   The first 10 inches of small intestine. Where all digestion is completed.

Ecosystem   Includes all the organisms in a given area as well as the abiotic (nonliving) factors with which they interact.

Ectoderm   The outermost layer of an embryo, which develops into skin and nervous system.

Egestion   Removal of metabolic waste.

Electron transport chain or **ETC**   Consists of a series of molecules within the cristae membrane of mitochondria that provides the energy to phosphorylate ADP into ATP during oxidative phosphorylation.

Endoderm   The innermost layer of an embryo, which develops into the viscera or the digestive system.

Endoplasmic reticulum   System of transport channels within the cytoplasm of a eukaryotic cell.

Endosperm   Food for the growing embryo in a monocot seed. The cells that make up the endosperm are triploid ($3n$).

Energy of activation   The amount of energy required to start a reaction.

Eohippus   A transition fossil that demonstrates that the ancient horse is an ancestor of the modern horse, *Equus*.

Epicotyl   Part of the embryo in a seed that becomes the upper part of the stem and leaves.

Epididymis   Part of testes where sperm become motile.

Epiglottis   Flap of cartilage in the back of the throat that directs food to the esophagus.

Erythrocytes   Red blood cells.

Ethylene   Gaseous plant hormone that promotes fruit ripening.

Eukaryotes   Cells that contain internal membranes. The opposite of prokaryotic cells.

Excited state   When an atom absorbs energy, its electrons move to a higher energy level.

Excretion   Removal of metabolic wastes.

Exocytosis   The release of substances from a cell.

Exons   Expressed sequences of DNA. DNA that codes for particular polypeptides.

Extremophiles   Organisms that live in extreme environments, like methanogens, halophiles, and thermophiles. These organisms make up the domain *Archaea.*

FAD or **flavin adenine dinucleotide**   Coenzyme that shuttles protons and electrons from glycolysis and the Krebs cycle to the electron transport chain.

Fallopian tube   *See oviduct.*

Fermentation   Anaerobic phase of cell respiration.

Filament   Threadlike structure that holds up the anther in the male part of a flower.

Filtration   Process that occurs in the nephron where nutrients and wastes diffuse from the glomerulus into Bowman's capsule.

Final transcript   The strand of mRNA that is sent to the ribosome after processing. The final transcript is much shorter than the initial transcript.

Fission   Division of an organism into two new cells. Reproduction in protists.

Fixed action pattern   Innate, highly stereotypical behavior, which once begun is continued to completion, no matter how useless or silly looking. FAPs are initiated by external stimuli called sign stimuli.

Food chain   Pathway along which food is transferred from one trophic or feeding level to another.

Founder effect   A small population, which is not representative of the larger population, breaks away from the larger one to colonize a new area. Rare alleles may be under- or overrepresented.

Fragmentation   A single parent organism breaks into parts that regenerate into new individuals. Reproduction in sponges, planaria, and sea stars.

Frameshift   An error in the DNA in which the entire reading frame is altered. This can be caused by an insertion or deletion.

G3P, glyceraldehyde-3-phosphate, or **PGAL**   First sugar produced by photosynthesis.

Gametangia   In primitive plants, a protective jacket of cells in which gametes and zygotes develop and which prevents drying out.

Gametophyte   Monoploid (*n*) generation of a plant.

Gastrin   Digestive hormone that stimulates sustained secretion of gastric juice from the stomach.

Gastrovascular cavity   Gastrocoel, primitive digestive cavity in hydra.

Gastrulation   The process by which a blastula develops into a gastrula with the formation of three embryonic layers.

Gel electrophoresis   Process that separates large molecules of DNA on the basis of their rate of movement through an agarose gel in an electric field.

Gene flow   Movement of alleles into or out of a population.

Genetic drift   Change in the gene pool due to chance. Two examples are the bottleneck effect and the founder effect.

Genetic engineering   Branch of science that uses recombinant DNA techniques for practical purposes, also called biotechnology.

Genome   An organism's genetic material. The human genome consists of 3 billion base pairs of DNA and about 30,000 genes.

Genotype   The kind of genes an organisms has.

Geographic isolation   Separation by mountain ranges, such as canyons, rivers, lakes, or glaciers, may cause significant isolation.

Gibberellins   Plant hormones that promote stem and leaf elongation.

Gizzard   Structure in birds, insects, and earthworms where mechanical digestion of food occurs.

Global warming   Increase in average temperature of Earth. It is due to the greenhouse effect.

Glucagon   Hormone released by the pancreas that raises blood sugar

Glycerol   Combines with fatty acids to make lipids.

Glycolysis   The anaerobic phase of aerobic respiration. One molecule of glucose breaks apart into two molecules of pyruvate.

Golgi apparatus   Cell organelle that packages and secretes substances for the cell.

Gradualism   The theory that organisms descended from a common ancestor gradually, over a long period of time, in a linear or branching fashion.

Grana   Membranes within chloroplasts that consist of thylakoid membranes and are the sites of the light-dependent reactions of photosynthesis.

Greenhouse effect   Carbon dioxide and water vapor in the air absorb much of the infrared radiation reflecting off Earth, causing the average temperature on Earth to rise.

Gross primary productivity   Amount of energy converted to chemical energy by photosynthesis per unit time in an ecosystem.

Ground state   Condition of an electron when it is not excited. It is in its lowest energy level.

Guard cells   Modified epithelium containing chloroplasts that control the opening and closing of the stomates by a change in shape.

Gymnosperms   Confers or cone-bearing trees.

Habituation   One of the simplest forms of learning in which an animal comes to ignore a persistent stimulus so it can go about its business.

Half-life   Amount of time it takes for a radioactive isotope to decay to half its mass.

Halophiles   Organisms that thrive in environments with high salt concentrations like Utah's Great Salt Lake.

Heat of vaporization   The amount of energy required to change a specified amount of liquid into a gas. Water has a high heat of vaporization.

Heliobacter pylori   Bacteria that is the cause of most ulcers.

Hemocoels   Sinuses. Cavities in the body of insects, like grasshoppers, for exchange of nutrients and wastes.

Hemophilia   An inherited disease caused by the absence of one or more proteins necessary for normal blood clotting.

Hermaphrodites   Organisms that contain both female and male sex organs.

Heterotroph hypothesis   This theory states that the first cells on Earth were anaerobic, heterotrophic prokaryotes.

Heterotrophs   Organisms that must take in all their nutrients.

Histamine   An important chemical in the immune system that triggers vasodilation (enlargement of blood vessels), which increases blood supply to an area. Histamine is also responsible for the symptoms of the common cold: sneezing, coughing, redness, itching and runny nose and eyes—all an attempt to rid the body of invaders.

Histones   Special proteins that wrap around DNA, forming chromatin network.

Homeostasis   Internal stability.

Homeotherm   Endotherm. Animals that maintain a consistent body temperature. Examples are birds, mammals, and some reptiles.

Homologous structures   The same internal bone structure, although the function of each varies. Examples of homologous structures are the wing of a bat, the lateral fin of a whale, and the human arm. If organisms have homologous structures, it means they have a common ancestor.

Huntington's disease   A degenerative inherited disease of the nervous system resulting in certain and early death. The gene that causes it is dominant, and onset is usually in middle age.

Hydrogen bonding   An intermolecular attraction between molecules that exert a strong pull on their electrons. This attraction keeps the two strands of a DNA molecule together.

Hydrophilic   Soluble in water. Hydrophilic substances are either polar or ionic.

Hydrophobic   Insoluble in water. Hydrophobic substances are nonpolar.

Hypertonic   Having greater concentration of solute than another solution.

Hypocotyl   Part of the embryo in a seed that becomes the lower part of the stem and the roots.

Hypothalamus   Major gland in the brain that is the bridge between the endocrine and nervous systems.

Hypotonic   Having less concentration of solute than another solution.

Imprinting   Learning that occurs during a sensitive or critical period in the early life of an individual and is irreversible for the length of that period.

Incomplete dominance   An inheritance pattern characterized by blending of traits. An example is crossing an animal with black fur with one with white fur, producing offspring with gray fur.

Ingestion   Intake of nutrients.

Initial transcript   Strand of mRNA before it is processed. The initial transcript is much longer than the final transcript.

Insertion A mutation where one nucleotide inserts itself into an existing strand. This mutation can cause a frameshift.

Insulin Hormone released by the pancreas that lowers blood sugar.

Intermolecular attraction Attraction between molecules. One example is hydrogen bonding.

Introns Intervening, noncoding sequences of DNA located between genes.

Inversion A chromosomal fragment breaks off and reattaches to its original chromosome but in the reverse orientation.

Ionic bonds Bonds between atoms that form by transferring electrons.

Irritability Ability to respond to stimuli.

Isotonic Solutions containing equal concentrations of solute.

Junk Noncoding regions of DNA. Most of the human genome consists of noncoding regions.

Karyotype Procedure that analyzes the size, shape, and number of chromosomes

Krebs cycle Also known as the citric acid cycle; the first stage of the aerobic phase of cellular respiration. It occurs in the inner matrix of mitochondria.

Lacteal Structures within the villi that line the small intestine and that absorb fatty acids and glycerol into the lymphatic system.

Lactic acid fermentation Occurs during strenuous exercise when the body cannot keep up with the increased demand for oxygen by skeletal muscles and pyruvic acid converts to lactic acid, which builds up in the muscle and causes fatigue and burning.

Law of dominance Mendel's first law that states that when two organisms, each homozygous (pure) for two opposing traits are crossed, the offspring will be hybrid but will exhibit only the dominant trait. The trait that remains hidden is the recessive trait.

Law of independent assortment Best demonstrated by the dihybrid cross. A cross that is carried out between two individuals hybrid for two or more traits that are not on the same chromosome—the resulting phenotype ratio is 9:3:3:1.

Law of segregation During the formation of gametes, the traits carried on homologous chromosomes separate.

Learning Sophisticated process in which the responses of the organism are modified as a result of experience.

Leucoplast Type of plastid that stores starch.

Leukocyte White blood cells.

Light-dependent reactions Part of photosynthesis that requires light, produces ATP, and releases oxygen.

Light-independent reactions Part of photosynthesis that does not require light directly, only the products of the light-dependent reactions. Sugar (PGAL) is the product.

Limiting factors Those factors that limit population growth. They are divided into two categories, density-dependent and density-independent factors.

Lipid One type of organic molecule. It consists of one glycerol plus three fatty acids.

Locomotion Moving from place to place.

Lysosome Cell organelle that consists of digestive (hydrolytic) enzymes and is the principal site of intracellular digestion in the cell.

Macroevolution Refers to speciation, the formation of an entirely new species.

Malpighian tubule Structures in grasshoppers for removal of the nitrogenous waste uric acid.

Malthus Published a treatise on population growth, disease, and famine in 1798 that influenced Darwin in the development of his theory of natural selection. Malthus stated that populations tend to grow exponentially, to overpopulate, and to exceed their resources.

Medusa Upside-down bowl-shaped body type.

Meiosis Type of cell division in sexually reproducing organisms that produces monoploid (*n*) gametes.

Menopause Cessation of the menstrual cycle.

Meristem tissue Plant tissue that is always dividing. An example is cambium tissue.

Mesoderm The middle layer of an embryo that develops into blood, bones, and muscle.

Mesoglea The middle layer of a two-layered animal, like sponges or hydra, which holds the two layers together.

Messenger RNA or **mRNA** Carries messages directly from DNA in the nucleus to the cytoplasm during protein synthesis.

Metabolism The sum total of all the life functions.

Methanogens   Organisms that obtain energy in a unique way by producing methane from hydrogen.

Microevolution   Changes in one gene pool of a population over generations.

Microfilaments   Made of the protein actin and help support the shape of the cell. They enable animal cells to form a cleavage furrow during cell division or the amoeba to move by sending out pseudopods.

Microtubules   Thick hollow tubes that make up the cilia, flagella, and spindle fibers.

Middle lamella   Layer of tissue between two cell walls of adjacent plant cells.

Mitochondrion   Cell organelle that produces ATP. Present in both plants and animals.

Mitosis   Type of cell division for growth and repair that produces two genetically identical daughter cells with the same chromosome number as the parent cell. Consists of four phases: prophase, metaphase, anaphase, and telophase.

Molecule   The name given to two or more atoms joined by a covalent bond.

Monocotyledon   Plant whose seed does not break into two parts. An example is corn.

Monohybrid cross   (*Tt* × *Tt*) A cross between two organisms that are each hybrid for one trait.

Monotremes   Egg-laying mammals, like the duck-billed platypus and the spiny anteater, which derive nutrients from a shelled egg.

mRNA   *See messenger RNA.*

Multiple alleles   When there are more than two allelic forms of a gene. For example, in humans, there are more than 2 alleles for blood type. There are A, B, and O.

Mutation   Any change in a gene or chromosome.

Mutualism   Symbiotic relationship where both organisms benefit (+/+). An example is the bacteria that live in the human intestine and that produce vitamins.

Mycorrhizae   Symbiotic structures consisting of the plant's roots intermingled with the hyphae (filaments) of a fungus, which greatly increase the quantity of nutrients that a plant can absorb.

Myosin   Myofilaments that make up the thick filaments in skeletal muscle.

NAD or **nicotinamide adenine dinucleotide**   Coenzyme that shuttle protons or electrons from glycolysis and the Krebs cycle to the electron transport chain.

Nematocysts   Stingers found in cnidocytes of cnidarians.

Nephridia   Structure in earthworms for excretion of the nitrogen waste urea.

Net primary productivity   Gross primary productivity minus the energy used by the primary producers for respiration.

Nondisjunction   An error that sometimes happens during meiosis in which homologous chromosomes fail to separate as they should.

Notochord   A rod that extends the length of the body and serves as a flexible axis. This is a characteristic of all chordates.

Nucleolus   Where components of ribosomes are synthesized. This is a prominent region within the nucleus of a cell that is not dividing.

Objective lens   The lens on a light microscope that is closest to the stage.

Ocular lens   Eyepiece of a microscope.

Omnivores   Animals that normally eat both meat and vegetables in their diet.

Oogenesis   Formation of ova by meiotic cell division.

Operant conditioning   Trial and error learning.

Organogenesis   Process by which cells continue to differentiate, producing organs from the three embryonic germ layers.

Origins of replication   Special sites where replication begins in eukaryotic cells.

Osmosis   Diffusion of water across a membrane.

Ovary   Swollen part of pistil of a flower that contains the ovule, where one or more ova are produced.

Oviduct or **fallopian tube**   Where fertilization occurs. After ovulation, the egg moves through the oviduct to the uterus.

Ovule   The structure within the ovary of a flower where the ova (female gametophyte) are produced.

Oxidation   Loss of electrons.

Oxidative phosphorylation   Process that provides most of the energy (ATP) produced during cell respiration.

Oxytocin   Hormone secreted by the posterior pituitary that stimulates the uterus and causes contractions during labor.

Pangaea   Single supercontinent on ancient Earth that slowly separated into seven separate continents over the course of 150 million years. This is evidence of the theory of continental drift.

Parallel evolution   Two related species that have made similar evolutionary adaptations after their divergence from a common ancestor.

Parasitism   Symbiotic relationship (+/−) where one organism, the parasite, benefits while the host is harmed.

Parenchyma cells   Traditional-looking plant cell. Have a primary cell wall that is thin and flexible but lack a secondary cell wall.

Parthenogenesis   The development of an egg without fertilization. The resulting adult is monoploid (haploid).

Pathogen   Organism that causes disease.

Pedigree   Family tree that indicates the phenotype of one trait being studied for every member of a family.

Peptidases   Enzymes that break down proteins into amino acids.

PGAL   *See G3P.*

Phagocytosis   Cellular process of engulfing food and encapsulating it in a vacuole.

Pharynx   Throat.

Phenotype   The traits an organism expresses.

Phenylketonuria   An inherited disease characterized by the inability to break down the amino acid phenylalanine. Requires elimination of phenylalanine from diet, otherwise serious mental retardation will result.

Photolysis   The process that occurs during the light-dependent reactions in which water is ripped apart to provide electrons to replace those lost by chlorophyll *a*. Oxygen is released.

Photosynthetic pigments   Chemicals that absorb light energy and use it to carry out photosynthesis. Examples are chlorophyll *a*, chlorophyll *b*, carotenoids, and phycobilins.

Phycobilins   A photosynthetic pigment.

Pinocytosis   Cellular process by which cells take in large dissolved molecules, referred to as cell drinking.

Pioneer organisms   The first to colonize a barren environment in primary ecological succession.

Pistils   Female part of the flower. Each pistil consists of an ovary, stigma, and style. Also called carpel.

Pith   Specialized region in the root of a plant for storage.

Plasma   Liquid portion of the blood.

Plasmodesmata   Openings in cell walls of plants for the passage of materials from one cell to another.

Plasmolysis   Cell shrinking, occurs when a cell is in a hypertonic environment.

Plastids   Organelles found only in plant cells; chloroplast is one example.

Poikilotherms   Cold-blooded. Examples are fish, amphibians, and reptiles.

Point mutation   A mutation in one nucleotide on DNA. Sickle cell anemia is caused by a point mutation.

Polarized   The condition of an axon of a nerve when it is at rest, also known as resting potential. Sodium and potassium are pumped to opposite sides of the membrane.

Polygenic There are more than two allelic forms of a gene. Examples include height or hair color in humans. The trait exhibits a multitude of variation.

Polymerase chain reaction   A cell-free, automated technique by which a piece of DNA can be rapidly copied or amplified. Useful in genetic engineering.

Polymers   Molecules that are chains of repeating units; proteins and DNA are examples.

Polyp   Vase-shaped body.

Polyploid   An organism with extra sets of chromosomes (3*n*, 4*n*, etc.). Commonly occurs in plants.

Polysaccharides   Molecules that consist of many monosaccharides joined together. Starch and chitin are examples.

Population   Group of individuals of one species living in one area that have the ability of interbreeding and interacting with each other.

Primary growth   Vertical growth of a plant.

Prions   Misfolded proteins that cause mad cow disease.

Prokaryotes   Cells that have no internal membranes or internal organelles, like nuclei or mitochondria.

Prostate gland   Large gland that secretes semen directly into the urethra.

Pseudocoelomate   An animal with a false coelom. An example is a roundworm.

Pseudopods   Means "false feet." This is how amoeba and white blood cells move from place to place.

Puberty   Onset of the menstrual cycle in girls and sperm production in boys.

Punctuated equilibrium   Theory that proposes that new species appear suddenly after long periods of stasis. Replaced gradualism theory in popularity.

Purines   The nucleotides adenine and guanine.

Pyloric sphincter   Band of muscle at the bottom of the stomach that keeps food in the stomach long enough to be digested.

Pyrimidines   Class of nucleotides, includes thymine and cytosine.

Radioisotopes   Radioactive isotopes, those that are decaying as they emit particles from the nucleus.

Reabsorption   In the nephron of the kidneys, a process by which most of the water and solutes (glucose, amino acids, and vitamins) that initially entered the tubule during filtration are transported back into the capillaries and, thus, back to the body.

Receptor-mediated endocytosis   Process by which cells take in specific molecules for which the cell has a specific receptor.

Recognition sequences or **sites**   The specific sites on DNA that restriction enzymes cut.

Recombinant DNA   The modern technique of taking DNA from two sources and combining them into one molecule or cell.

Reduction   Gain of electrons.

Reduction division   Another name for meiosis I, the division where homologous pairs separate.

Regulation   Ability to maintain internal stability, homeostasis.

Releaser   Sign stimuli exchanged between members of the same species.

Replication bubbles   Sections of DNA where the two strands separate in order to enable replication to occur rapidly. There are thousands of these bubbles along the DNA molecule, which speed up the process of replication along the giant human DNA molecule.

Replication fork   A Y-shaped region where the new strands of DNA are elongating during DNA replication.

Reproduction   Ability to generate offspring.

Resolution   A measure of clarity of an image seen under a microscope.

Respiration   Metabolic processes that produce energy (adenosine triphosphate or ATP) for all the life processes.

Restriction enzymes   Extracted from bacteria; they cut DNA at specific recognition sequences or sites, such as GAATTC.

Restriction fragments   The fragments of DNA that result from cuts made by restriction enzymes.

Rhizobium   Symbiotic bacterium that lives in the nodules on roots of specific legumes and that fixes nitrogen gas from the air into a form of nitrogen the plant requires.

Ribosomal RNA or **rRNA**   Structural RNA that is synthesized in the nucleolus. Along with proteins, it makes up the ribosome.

Ribosome   Structure in cells where proteins are synthesized.

RNA processing   Occurs in the nucleus before the newly formed mRNA strand is sent out to the ribosome.

rRNA   *See ribosomal RNA.*

Saprobes   Organisms that obtain food from decaying organic matter.

Sclerenchyma cells   Plant cells that have very thick primary and secondary cell walls fortified with lignin. Their function is purely support.

Scrotum   Sac outside the abdominal cavity that holds the testes. The cooler temperature there enables sperm to survive.

Secondary growth   Lateral growth of a plant.

Secretin   Digestive hormone that stimulates the pancreas to release bicarbonate to neutralize acid in duodenum.

Secretion   Process that occurs in the tubule of the nephron, which is the active, selective uptake of molecules that did not get filtered into Bowman's capsule.

Semiconservative replication   The way in which DNA replicates itself. The new DNA molecule consists of one old strand and one new strand.

Sepals   Outermost circle of leaves around a flower that are green and closely resemble ordinary leaves.

Serum   Plasma without the clotting factors.

Sessile   Nonmoving.

Sex-influenced trait   Inheritance is influenced by the sex of the individual carrying the traits.

Sex-linked Traits carried on the X chromosome.

Sieve tube elements   Make up phloem, along with companion cells.

Sign stimuli   Initiate a fixed action pattern.

Sinoatrial (SA) node   Pacemaker of the heart.

Sinuses   Cavities in the body for exchange of fluid. Called hemocoels in grasshoppers.

Sister chromatids   A replicated chromosome consists of two of these, where one is an exact copy of the other.

Spliceosomes   Special molecules that assist in the editing of mRNA during RNA processing.

Sodium-potassium pump   Elaborate molecule that sits within neuron membranes and pumps sodium and potassium ions across the membrane.

Solute   Substance that is dissolved in a solvent.

Solvent   Substance that does the dissolving. In a solution of salt and water, the salt is the solute and the water is the solvent.

Somatic system   Part of the nervous system that controls voluntary muscles.

Species   A population whose members have the potential to interbreed in nature and produce viable, fertile offspring.

Specific heat   Amount of heat that must be absorbed in order for 1 gram of a substance to change its temperature by 1° Celsius.

Spermatogenesis   Formation of sperm by meiotic cell division.

Spindle fibers   Made of microtubules, these assist in cell division.

Spontaneous generation   The theory that living things emerge from nonliving or inanimate objects.

Sporophyte   The diploid ($2n$) generation of a plant.

Sporopollenin   Tough polymer that is resistant to almost all kinds of environmental damage and that protects plants in a harsh terrestrial environment. It is found in the walls of spores and pollen from which it gets its name.

Stabilizing selection   This type of natural selection eliminates the extremes and favors the more common intermediate forms.

Stamen   Male part of the flower, consists of anther and filament.

Stele   Vascular cylinder in a plant root.

Stomates   Openings in leaves to exchange photosynthetic gases: water vapor, carbon dioxide, and oxygen.

Stroma   Part of the chloroplasts that holds the grana.

Style   Long, usually thin stalk of the pistil of a flower.

Substrate level phosphorylation   Process by which ATP is produced as a special enzyme moves a phosphate from one molecule to ADP. How energy is produced during glycolysis and the Krebs cycle.

Symplast   System of transport within a plant consisting of openings in cell walls called plasmodesmata.

Synapsis   The process in which homologous chromosomes pair up. This occurs during prophase I.

Synthesis   Combining of small molecules or substances into larger, more complex ones.

Systole   Contraction of the ventricles of the heart. Normal systolic pressure is 120 mm Hg.

Taproot   Single, large root like a carrot.

Taxon (Taxa, pl.)   Levels of organization in our system of classification: domain, kingdom, phylum, class, order, family, genus, and species. (The book you are using may not include the newest system that includes "domain.") Domain or kingdom includes the most different organisms, while species includes the most similar organisms.

Taxonomy   System by which we name and classify all organisms, living and extinct.

Tay-Sachs disease   An inherited disease with onset early in life and that is caused by lack of the enzyme necessary to break down lipids necessary for normal brain function. It is common in Ashkenazi Jews and results in seizures, blindness, and early death.

Telomeres   Special nonsense nucleotide sequences (TTAGGG) located at the ends of chromosomes that repeat thousands of times. These ends protect the DNA during cell division.

Territory   Area an organism defends and from which other members of the community are excluded.

Testcross   Backcross. A technique to determine whether an individual plant or animal showing the dominant trait is homozygous dominant (*BB*) or hybrid (*Bb*). The organism of unknown genotype is crossed with a recessive individual.

Testes (testis, sing.)   Male gonads; the site of sperm formation.

Theory of endosymbiosis   This theory states that cell organelles, like mitochondria, were once tiny, free-living prokaryotic organisms that took up permanent residence inside larger prokaryotic organisms.

Thermophiles   Organisms that thrive in very high temperatures, like in the hot springs in Yellowstone Park or in deep-sea hydrothermal vents.

Thrombocytes   Platelets.

Thylakoids   Specialized membranes that make up the grana in chloroplasts, the site of the light-dependent reactions.

Tracer   Radioactive substance that can be used to track a substance as it moves through an organism or through a metabolic pathway. They can be used in research or as a diagnostic tool in medicine.

Tracheids   Cells that, along with vessel elements, make up xylem.

Transcription   The process by which DNA makes RNA.

Transfer RNA or tRNA   Shaped like a cloverleaf and carries amino acids to the mRNA at the ribosome as proteins are synthesized.

Transformation   A phenomenon in bacteria. They have the ability to transform themselves by transferring genetic factors from one bacteria cell to another.

Translation   The process in which the DNA code is translated into an amino acid sequence and a polypeptide is formed. Occurs at the ribosome.

Translocation   A fragment of a chromosome becomes attached to a nonhomologous chromosome.

Transpiration   Loss of water from a leaf of a plant through stomates.

Transport   Intake and distribution of substances in cells or tissue, not the same thing as locomotion.

Trichomes   Tiny, spikelike projections on some leaves for protection.

Triploblastic   An animal consisting of three cell layers. This includes every animal more sophisticated than flatworms.

Trisomy   Having a chromosome in triplicate instead of duplicate. Down syndrome is cause by trisomy of the 21st chromosome.

tRNA   *See transfer RNA.*

Tropic hormones   Hormones that stimulate other glands to release their hormones.

Tropism   Growth of a plant toward or away from a stimulus.

Turgid   A property of plant cells, swollen.

Turgor pressure   Pressure exerted when a plant cell swells.

Ultracentrifuge   A machine that spins mashed tissue so quickly that it separates the homogenate into separate pellets of different organelles.

Urethra   Tube that carries semen and urine in males. In females, it carries only urine.

Uterus   Where the blastula stage of the embryo implants and develops during the nine-month gestation if fertilization occurs.

Vacuoles   Organelles in cells whose function is storage.

Vagina   Birth canal.

Vas deferens   Duct that carries sperm during ejaculation from the testes to the penis.

Vegetative propagation   Asexual reproduction in a plant where a piece of the root, stem, or leaf produces an entirely new plant genetically identical to the parent plant. Examples are grafting, cuttings, bulbs, and runners.

Vesicles   Small vacuoles.

Vessel elements   Structures that, along with tracheids, make up xylem.

Vestigial structures   Structures that are remnants of an earlier active structure, such as the appendix. They are evidence that animals have evolved.

Villus (villi, pl.)   Millions of fingerlike projections that line the small intestine and absorb all nutrients that were previously released from digested food.

Visible spectrum   Wavelengths of light that humans can see: 380 nm to 750 nm.

Wave of depolarization   The condition of an axon when an impulse is passing, also referred to as an action potential.

Xylem   Vessels in plants that carry water and nutrients from the soil to the rest of the plant.

Yolk sac   Membrane that encloses the yolk of an egg; food for the growing embryo.

Bibliography

Arms, Karen and Pamela S. Camp. *Biology*, 4th edition. Fort Worth, TX: Harcourt Brace and Co., 1995.

Campbell, Neil A. and Jane Reece. *Biology*, 6th edition. Menlo Park, CA: Addison Wesley Longman, 1999.

Curtis, Helena and N. Sue Barnes. *Biology*, 5th edition. New York: W.H. Freeman, 1990.

Gould, James L. and William T. Keeton. *Biological Science*, 6th edition. New York: W.W. Norton, 1996.

Guttman, Burton S. *Biology*. New York: McGraw-Hill Co., 1999.

Interview with Bernard Wides of American Museum of Natural History, NY. August 11, 2006.

Klug, William S. and Michael R. Cummins. *Concepts of Genetics*. 6th edition. Upper Saddle River, NJ: Prentice Hall, 2000.

Krogh, David. *Biology, A Guide to the Natural World*. Upper Saddle River, NJ: Prentice Hall, 2000.

Mader, Sylvia S. *Biology*, 7th edition. Dubuque, IA: McGraw Hill, 2001.

Matthews, Christopher K. and K. E. van Holde. *Biochemistry*. Redwood City, CA.: Benjamin/Cummings Publishing Co., 1990.

McFadden, C. and W. Keeton. *Biology—Exploration of Life*. New York: Norton, 1995.

Miller, Kenneth R. and Joseph Levine. *Biology*. Upper Saddle River, NJ: Prentice Hall, 2003.

Miller, Tyler G. *Living in the Environment*. Pacific Grove, CA: Brooks/Cole, 2000.

Morgan, J. and M. Carter. *Investigating Biology*, 2nd edition. Redwood, CA: Benjamin/Cummings, 1996.

Purves, William K., Gordon H. Orians, and H. Craig Heller. *Life: The Science of Biology*, 6th edition. New York: W.H. Freeman, 2000.

Raven, Peter H. and George B. Johnson, *Biology*, 5th edition. Dubuque, IA: William C. Brown/McGraw Hill Publishers, 1999.

Solomon, Eldra, Linda R. Berg, and Diane W. Martin. *Biology*, 5th edition. Harcourt College Publishers, 1999.

Wallace, Robert A., Gerald P. Sanders, and Robert J. Ferl. *Biology: The Science of Life*, 4th edition. New York: Addison-Wesley Publishing Co., 1996.

Index

SAT BIOLOGY CD-ROM Minimum Requirements

The following documentation applies to the *SAT Subject Test Biology E/M* book with CD-ROM. Please disregard this information if your version does not contain the CD-ROM available in September 2007.

Documentation

MINIMUM HARDWARE REQUIREMENTS

The program will run on a PC with

1. Intel Pentium® 166 MHz or equivalent processor
2. 32 MB RAM
3. MS Windows® 95/98/NT/2000/ XP
4. SVGA (256 Colors) Monitor
5. 8X CD-ROM drive
6. Keyboard, Mouse

The program will run on a Macintosh® with

1. PowerPC 8600
2. Operating System 9.0 (or higher)
3. 32 MB RAM
4. SVGA (256 Colors) Monitor
5. 8X CD-ROM drive
6. Keyboard, Mouse

LAUNCHING THE APPLICATION

Barron's SAT Biology CD-ROM includes an "autorun" feature that automatically launches the application when the CD is inserted into the CD drive. In the unlikely event that the autorun features are disabled, alternate launching instructions are provided below.

Launching instructions for the PC

Windows® 95/98/NT/2000/XP:

1. Put the Barron's SAT Biology CD-ROM into the CD-ROM drive.
2. Click on the Start button and choose Run.
3. Type *D:\SATBiology.EXE* (assuming the CD is in drive D), then click OK.

Launching instructions for Macintosh®

1. Put the Barron's SAT Biology CD-ROM into the CD-ROM drive.
2. Double-click the SAT Biology Installer icon.
3. Follow the onscreen instructions.